Sedimentology Review/1

Sedimentology Review/1

EDITED BY V. PAUL WRIGHT

Editorial Board: Members of the
Postgraduate Research Institute for Sedimentology
University of Reading

OXFORD

BLACKWELL SCIENTIFIC PUBLICATIONS

LONDON EDINBURGH BOSTON

MELBOURNE PARIS BERLIN VIENNA

© 1993 by
Blackwell Scientific Publications
Editorial Offices:
Osney Mead, Oxford OX2 0EL
25 John Street, London WC1N 2BL
23 Ainslie Place, Edinburgh EH3 6AJ
238 Main Street, Cambridge
  Massachusetts 02142, USA
54 University Street, Carlton
  Victoria 3053, Australia

Other Editorial Offices:
Librairie Arnette SA
2, rue Casimir-Delavigne
75006 Paris
France

Blackwell Wissenschafts-Verlag
Meinekestrasse 4
D-1000 Berlin 15
Germany

Blackwell MZV
Feldgasse 13
A-1238 Wien
Austria

First published 1993

DISTRIBUTORS

Marston Book Services Ltd
PO Box 87
Oxford OX2 0DT
(*Orders*: Tel: 0865 791155
         Fax: 0865 791927
         Telex: 837515)

USA
Blackwell Scientific Publications, Inc.
238 Main Street
Cambridge, MA 02142
(*Orders*: Tel: 800 759-6102
           617 876-7000)

Canada
Oxford University Press
70 Wynford Drive
Don Mills
Ontario M3C 1J9
(*Orders*: Tel: 416 441-2941)

Australia
Blackwell Scientific Publications
Pty Ltd
54 University Street
Carlton, Victoria 3053
(*Orders*: Tel: 03 347-5552)

A catalogue record for this title
is available from the British Library

ISBN 0-632-03102-6

ISSN 0967-8883

Contents

List of contributors

Editor

V.P.WRIGHT *Postgraduate Research Institute for Sedimentology, University of Reading, PO Box 227, Whiteknights, Reading RG6 2AB, UK*

Contributors

J.R.L.ALLEN *Postgraduate Research Institute for Sedimentology, University of Reading, PO Box 227, Whiteknights, Reading RG6 2AB, UK*

D.J.BOTTJER *Department of Geological Sciences, University of Southern California, University Park, Los Angeles, CA 90089-0740, USA*

R.J.CHEEL *Sedimentary Geology Research Group, Department of Geological Sciences, Brock University, St Catharines, Ontario L2S 3AI, Canada*

I.J.FAIRCHILD *School of Earth Sciences, University of Birmingham, Edgbaston, Birmingham B15 2TT, UK*

L.A.FRAKES *Department of Geology and Geophysics, University of Adelaide, PO Box 498, Adelaide, South Australia 5001, Australia*

J.E.FRANCIS *Department of Earth Sciences, The University of Leeds, Leeds LS2 9JT, UK*

D.A.LECKIE *Institute of Sedimentary and Petroleum Geology, Geological Survey of Canada, 3303–33rd Street NW, Calgary, Alberta T2L 2A7, Canada*

D.J.ROSS *Sun International Exploration and Production Company Ltd, Sun Oil House, 80 Hammersmith Road, London W14 8YS, UK*

C.E.SAVRDA *Department of Geology, Auburn University, Auburn, AL 36849-5305, USA*

P.W.SKELTON *Department of Earth Sciences, The Open University, Walton Hall, Milton Keynes MK7 6AA, UK*

M.E.TUCKER *Department of Geological Sciences, University of Durham, Durham DH1 3LE, UK*

G.P.WEEDON *School of Geological and Environmental Sciences, Luton College of Higher Education, Park Square, Luton, Bedfordshire LU1 3JU, UK*

Preface

The pace of new developments in sedimentology never seems to slow down. Not only is the discipline entering into a new phase of synthesis, provided by the concepts of sequence stratigraphy and the awareness of the importance of orbital-forcing in influencing facies sequences, but the application of sedimentology to issues such as environmental management and environmental change offers new avenues of research. All these aspects are reflected in this first volume.

As the discipline matures, its scope broadens and this series is an attempt to help provide a means of communication between the various branches of sedimentology. Few of us are able to keep abreast with developments outside our particular research field and these short reviews are designed to allow us to follow a range of such developments.

Their specific role is to provide a set of up-to-date reviews for the student and the practising sedimentologist. Authors have been requested to provide key references in their chapters. Not all the chapters are general reviews. There are many unusual sedimentary sequences in the rock record and this series intends to bring some of the lesser known ones to the attention of the community. The first chapter in this volume describes just such an example. Ian Fairchild reviews the highly unusual occurrence of carbonates and glacial deposits which are widespread in late Precambrian sequences.

The growing interest in palaeoclimatology is reflected in the second chapter, by Jane Francis and Larry Frakes. A great deal of research has been carried out on the distinctive global climatic regime of the Cretaceous and this chapter synthesizes the sedimentological and palaeontological evidence.

In the next chapter, Graham Weedon reviews the problems associated with recognizing orbitally-forced cycles in the geological record. The interest in this theme is enormous with the potential that such cycles may eventually become one of the key foundations for lithostratigraphy and even chronostratigraphy.

Continuing the theme of sea-level and climatic controls, Maurice Tucker offers a stimulating review on the relationships between systems tracts, climate and diagenesis in carbonate sequences.

Many sedimentologists are rediscovering the importance of fossils and indeed many palaeontologists are now integrating more sophisticated facies analysis into their studies. In view of this, two chapters concentrating on aspects of sedimentary palaeobiology are included in this volume. Chapter 5, by Donald Ross and Peter Skelton on rudist accumulations, addresses the misconceptions held by many of us regarding these extinct bivalves. Such revisionist research will be an important aspect of future issues. Chapter 6, by David Bottjer and Charles Savrda, is a review of oxygen-related mudrock biofacies. Mudrocks are, volumetrically, the most important sediment type, and palaeobiological and geochemical evidence is crucial to their interpretation. Both these chapters also relate to the economic dimension of our subject. Rudist build-ups form major reservoirs in the Middle East, the United States and Central America, while mudrocks provide most of the world's oil-prone source rocks.

The last two chapters discuss aspects of physical sedimentology and facies analysis. Chapter 7, by Rick Cheel and Dale Leckie, provides a review on hummocky cross-stratification (HCS). Rarely, if ever, has a sedimentary structure attracted such interest as has HCS. It is a sobering thought that such an important and controversial feature was not recognized properly until the mid-1970s. In the last chapter, John Allen provides an introduction to one of the least appreciated (geologically-speaking) sedimentary systems, namely estuaries. There are still surprisingly few records of ancient analogues in the geological record and this review will do much to introduce the reader to the basic processes operating in such complex systems.

On behalf of the Editorial Board, I thank the authors for their excellent contributions and hope that the readers find these reviews useful and stimulating. The patience and support of Simon Rallison of Blackwell Scientific Publications was much appreciated. Thanks go also to the many reviewers for their efforts.

V. Paul Wright

1 Balmy shores and icy wastes: the paradox of carbonates associated with glacial deposits in Neoproterozoic times

IAN J. FAIRCHILD

Introduction

The extraordinary events of Neoproterozoic times, 1000 to 570 Ma, have been slow to register in the consciousness of most western earth scientists, not least because the sedimentary record of that era is best displayed in regions remote from their centres of civilization. There is currently much interest in the patterns of evolution of primitive organisms prior to the Cambrian radiations and changes in atmospheric chemistry, both of which are quite possibly linked to the bizarre palaeoclimatology of those times. It was Brian Harland (1964), who first clearly stated how important the issue might be, by developing the concept that glaciation had been so severe that major ice sheets reached sea-level even at low latitudes. Part of the evidence for this was the presence of allegedly glacial sediments interbedded with carbonates, often dolomitic in composition, and inferred to be warm-water in origin. The origin of these carbonates is the subject of this review.

There are many implications of Harland's hypothesis, and one that he pursued vigorously was that, given such intense glaciation, individual glacial formations would be widespread and hence could be used as a primary tool of late Precambrian chronostratigraphy. Others objected however, and Schermerhorn (1974) developed an alternative framework in which the supposed tillites were reinterpreted as tectonically-generated sediment-gravity flows in the context of rifting basins. Palaeomagnetists repeatedly found evidence for low-latitude glaciation, but early records were mostly demolished as spurious or secondary magnetizations. Williams (1975) accepted both the glacial interpretation and low palaeolatitudes, but proposed dramatic *seasonal* shifts in temperature as a consequence of increased obliquity (tilt of the Earth's spin axis) in those times.

In part, these arguments were fuelled by assertions about the genesis both of the alleged glacial deposits and their associated carbonates, but based on a limited range of observations. In the last 20 years, specifically sedimentological approaches to these questions have transformed the nature of the arguments: for example we now can identify glacigenic strata with considerable certainty. Also the processes of formation of the carbonates, and hence their likely palaeoclimatic setting can now be discussed more confidently, and in detail.

Figure 1.1 illustrates the stratigraphic context of the occurrences of carbonates* with glacial deposits. The labels 'pre-glacial', 'glacial' and 'post-glacial' emphasize the point that the carbonate rocks fall into two groups: those within and those outside of glacigenic formations. This chapter introduces the Neoproterozoic sections and discusses the historically-important arguments, reviews the nature of the bounding carbonates and discusses the relationships they display with the glacial units. The nature of the carbonate within glacial units, both detrital and authigenic, is then outlined — thereby allowing us to review the concept of low-latitude glaciation and finally to look forward to the imminent exciting period of intensive research on Neoproterozoic sections in which some of the outstanding problems may be resolved.

The Neoproterozoic successions

Precambrian deposits of <1000 Ma are present on

---

* The term 'carbonates' is used to refer to limestone/calcite and/or dolostone/dolomite where the distinction between these species is unimportant. Nearly all the rocks described in this chapter contain significant proportions of carbonate minerals. The terms 'sandstone' and 'shale' are used to emphasize the siliciclastic content of some lithologies, whereas the terms mudrock, arenite, rudite, and diamictite (poorly-sorted sedimentary rock including a gravel component) are used simply to refer to grain size. The use of the 'dol-' prefix denotes a dolostone (e.g. dolarenite is composed of sand-sized clasts of dolomite).

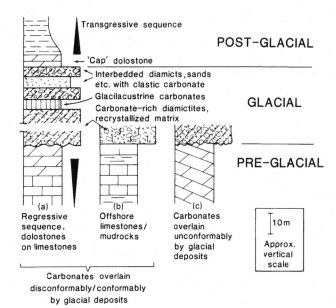

Fig. 1.1 A summary of the stratigraphic relationships of carbonates with Neoproterozoic glacial deposits. Note the interpreted sharp limits of glacial conditions. The 'cap' dolostone (dolomite) is a very widespread phenomenon (see text) as is the presence of clastic dolostone (and sometimes limestone) in glacial sediments. The relationship with underlying carbonates shown in (a) is best documented from the North Atlantic area; (b) is known, for example, in East Greenland and North America; (c) is common in, for example, Australia, West Africa and Australia.

all present-day continents. They vary from flat-lying unmetamorphosed cratonic-cover sequences to high-grade metasediments, although the most common successions are folded and very weakly metamorphosed. Conformable successions representing at least 100 Ma range in thickness from a few hundred metres to several kilometres.

Glacial sediments may well be represented on all the continents now that a possible example from Antarctica has been found, although they only occupy a subordinate part of any individual succession. The major single source of information on the glacial deposits is the compilation of Hambrey & Harland (1981). Analysis of these data confirms that three major periods of glaciation can be recognized around 900 Ma (Lower Congo), 800 Ma (Sturtian) and 650 Ma (Varangian). Commonly, more than one distinct glacial unit occurs within a glacial period; South Australia has a particularly full set of glacial events. Two Varangian glacial units (Figs 1.2–1.4) are the norm in the North Atlantic region towards which this review is biased.

Carbonate rocks, although not universal, are found in a significant number of the sections, and their co-occurrence with glacial deposits has long been recognized as a peculiar feature of late Proterozoic as opposed to other glacial periods. Carbonates are particularly abundant in association with the Varangian glacial episodes. Studies of geochemistry and textures indicate that virtually no burial

Fig. 1.2 Kap Weber, central East Greenland (Herrington & Fairchild, 1989). Slope shales and carbonates (1) overlain by platform carbonates (2, about 70 m thick), then aeolian sandstones (3) and diamictites of the first Varangian glaciation (Moncrieff & Hambrey, 1990).

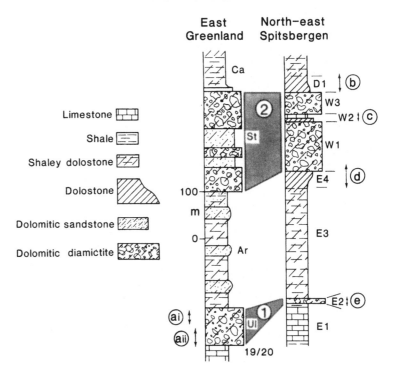

Fig. 1.3 The two glacial units (labelled 1 and 2) of the East Greenland–NE Svalbard basin. Letters a to e mark the stratigraphic positions of the enlarged sections shown in Fig. 1.4. Ca, Canyon; St, Storeelv; Ar, Arenean; Ul, Ulveso; 19/20, 'bed-groups' 19 and 20; Dl, Dracosien member 1; W1–W3, Wilsonbreem members 1–3; E1–E4, Elbobreen members 1–4.

diagenetic alteration is present in some sections, and even in greenschist facies rocks stratigraphic relationships are still well preserved.

Earlier ideas on the origin of the carbonates

In publications up to around 1975, there was little sedimentological detail in the references to carbonates, but an extensive range of hypotheses for their origin had been proposed. There was considerable emphasis (Spencer, 1971) on the implications of warm temperature from the occurrence of dolostones, but whilst some used this to argue against glaciation (e.g. Schermerhorn, 1974), others took it that major climatic shifts occurred in response either to rapid continental drift or global temperature changes. However, although dolomitization is undoubtedly favoured by increasing temperatures, we now know from modern examples that dolomite can form under cool conditions.

The occurrence of stromatolites within and outside glacial units was also commonly taken as a warm-water indicator (e.g. Williams, 1975), but some of those inside glacial formations were misidentified (e.g. Spencer, 1971). These papers were written before cold-water Antarctican glaciolacustrine stromatolites became known.

A third general criterion for warm conditions was the presence of reddened sediments, including diamictites which were often taken to reflect the erosion of pre-existing lateritic soils formed either before the glaciation, during a hot interglacial period, or seasonally during a glacial period. Following better understanding of secondary reddening processes, it is now more logical to discard any climatic significance to reddening. Indeed, many glacial sediments, rich in unweathered minerals and low in organic matter, are prime material for the development of secondary haematite, and supporting evidence in the presence of Liesegang banding on various scales has been recognized (Fairchild & Hambrey, 1984). After 1975, arguments for warm conditions were made on the basis of a variety of sedimentary criteria which will be examined in a later section.

Whilst it was recognized that much carbonate within glacial formations could be clastic in origin (Harland & Herod, 1975), the presence of bedded, apparently fine-grained carbonate beds led some to favour cold-water mechanisms of formation. Carey & Ahmad (1961) postulated the existence of carbonate formation from brines beneath ice shelves as a consequence of seawater freezing onto cold glacier ice; this proved a popular theory. However, it turns

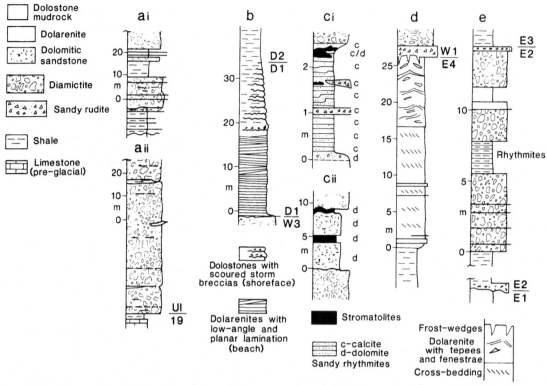

Fig. 1.4 Examples of lithologic logs in key parts of the sections shown in Fig 1.3. All lithologies are carbonate-rich. (a) Stratal sequences resulting from processes of melt-out, current reworking and gravity-controlled resedimentation in a glacimarine environment. Detrital dolostone is abundant, but subordinate to detrital limestone in (a)ii (Storeelv locality; modified from Moncrieff & Hambrey, 1990). (b) A cap dolomite of clearly transgressive character resting on dolomitic diamictite (Ditlovtoppen locality; expanded from Fairchild & Hambrey, 1984). (c) Complex glaciolacustrine facies: (c)i illustrates an advance–retreat cycle in dominantly calcitic rhythmites and stromatolites (freshwater lake), (c)ii shows dolomitic, ^{18}O-rich saline lake facies (Fairchild et al., 1989). (d) Regressive sequence of warm-climate dolostones cut by frost wedges and overlain by terrestrial glacial sediments (Fairchild & Hambrey, 1984). (e) Section through the thin representative of the earlier glaciation in north-east Spitsbergen dominated by clastic glacimarine dolostones in which syn-depositional recrystallization has occurred (Dracosien locality; Fairchild & Hambrey, 1984; Fairchild et al., 1989).

out that seawater freezing seems to be of more restricted occurrence than originally envisaged and the postulated brines have not been found in modern settings. Indeed melting, rather than freezing, seaward of the grounding line is the most common process, and this phenomenon provides the basis for a widely-used sedimentation model for many of the Late Proterozoic diamictites (e.g. Moncrieff & Hambrey, 1990). Carey & Ahmad (1961) also linked their model to the formation of glendonites, crystal pseudomorphs common in sediments associated with the Upper Palaeozoic Gondwana glacials. Glendonites are now known to be pseudomorphs after the generally cold-water carbonate ikaite

$(CaCO_3 \cdot 6H_2O)$. Distinctive pseudomorphs occur around the base of the Varangian glacial Port Askaig Formation (Spencer, 1971), but it is not known whether they were originally ikaite or gypsum.

Another category of hypothesis relates the origin of thin (metre-scale) cap dolomites to the warming of ocean water on deglaciation because of the decreasing solubility of carbonates with increasing temperature. However, the mechanism is quantitatively implausible.

Since 1975, increasing knowledge of high-latitude skeletal carbonates led some to postulate an analogy with Precambrian carbonates associated with glacial deposits (Bjørlykke et al., 1978), but since the controls on the precipitation of skeletal and non-skeletal carbonates are very different, this analogy is of limited usefulness.

In summary, the palaeoclimatic criteria developed in earlier work were based on too generalized observations to be diagnostic, and the modern analogues sought often seemed to have flaws. In the following section, I show how consideration of the sedimentological context in more detail sheds light on the problem.

'Normal' Neoproterozoic carbonates

General

We need to consider the characteristics of carbonate rocks that are not associated with glacial strata as a starting point. Precambrian carbonate rocks present a number of peculiarities in comparison to Phanerozoic ones, not just in the absence of skeletal organisms, but also in the abundance of stromatolites and subaqueous shrinkage cracks. A marine rather than lacustrine origin for the carbonates is difficult to establish categorically everywhere, but certain widespread microfossils are good evidence of marine conditions (Knoll, 1985), and distinct lacustrine intervals contrast with most of the carbonate facies (Fairchild & Hambrey, 1984). Figure 1.5 illustrates schematically a number of the common facies types to serve as a basis for discussion.

Carbonate slope facies (Fig. 1.5a,b) are not often preserved, but are distinctive. Being characterized by slumping and sediment-gravity flow deposits, there is potential for confusion with glacimarine deposits, but in practice the local origin of clasts by reworking of syn-depositionally lithified material, and the absence of coarse siliciclastic debris and dropstones allow confusion to be avoided. In examples found in East Greenland, the Sr-rich nature of slumped limestones with nodules or shrinkage-crack fill cements implies either lithification *of* 'periplatform' aragonite ooze, or lithification *by* aragonite.

In shelf facies too, an aragonitic precursor to limestone is often indicated (e.g. Tucker, 1986), although primary Mg calcite has been identified in the situation of Fig. 1.5i, and it is difficult to rule out a mixed primary mineralogy for many micrites. Arguably bimineralic aragonite–Mg calcite ooids were common. The stabilized limestones may lack primary structures, although lamination, discrete storm beds, and limestone–shale cycles are present

in some cases, in parallel with Phanerozoic facies. In Precambrian times, given the absence of calcareous plankton, precipitation of carbonate mud is likely to have been stimulated by primary productivity or evaporation. Microspar-cemented shrinkage cracks (lacking sediment because they were not open to the sediment surface) are widespread (Fig. 1.5c,j,m) in both apparently open-shelf and lagoonal sediments. Origin of cracks by synaeresis, specifically shrinkage induced by salinity changes, is plausible but unproven: if valid, significant evaporation is implied.

Columnar stromatolite bioherms and biostromes (Fig. 1.5f,i,k) are common in subtidal facies, and accreted by a combination of trapping of carbonate ooze and siliciclastic sediment, and *in situ* precipitation, although it is difficult to quantify the relative contribution of the two process. The stability of the environments often led to distinctive morphologies being developed which have been classified using the Linnean system of taxonomy. In deeper waters (Fig. 1.5c (base), h) in which carbonate was either not being precipitated, or else had been dissolved, organic remains of microbial mats are found.

Shoal facies often contain coated grains up to several millimetres in size (Fig. 1.5l) which has been taken as reflecting high aqueous supersaturations for calcium carbonate. They can be reliably distinguished from oncoids, which are often bean-shaped and smaller.

Lagoonal facies (Fig. 1.5e,m,n) commonly display shrinkage cracks, filled by cement or sediment, as well as evidence of storm activity. Upwardly-shallowing cycles to tidal flat sediments (Fig. 1.5d,o,q), either to sandflats with dolostone intraclasts or to microbial laminites, are present. Anhydrite relics and pseudomorphs have been discovered in several examples.

Distinctive oolitic supratidal facies (Fig. 1.5p) with evidence of extensive syn-sedimentary cementation associated with cement-lined fractures, and tepee structures, can be compared with modern Australian examples of cementation by seepage of saline groundwater or seawater. Associated calcified erect stalked cyanobacteria are closely similar to modern supratidal types in the Bahamas.

The mineralogy referred to on Fig. 1.5, excludes burial diagenetic effects, such as epigenetic dolomitization. This reveals the interesting generality that the shallowest-water facies are generally composed completely of dolomite, whereas shallow subtidal facies are variably dolomitized, and outer-shelf

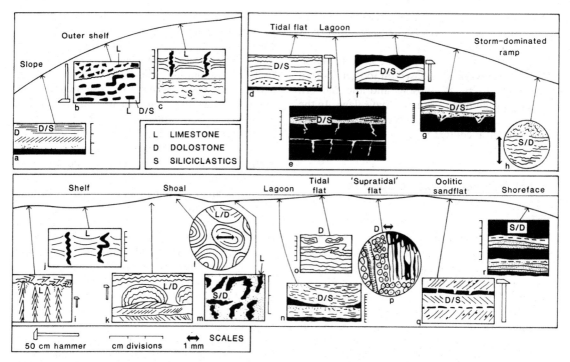

Fig. 1.5 Schematic summary of Neoproterozoic carbonate facies outside of glacial formations: (a) turbidites derived from platform; (b) syn-sedimentary lithification, slumping and breccia formation; (c) cemented shrinkage cracks (now compacted); siliciclastics with microbial mats; (d) tidal flat stromatolites building on flake breccia; (e) wave-rippled storm sands and sediment-filled shrinkage cracks in lagoonal muds; (f) sediment-rich smooth stromatolitic bioherms; (g) storm beds with gutters eroding underlying dolostone crusts; (h) siliciclastics with microbial mats interrupted by thin storm layer; (i) biostrome of conical stromatolites truncated by storm damage; (j) limestone with cemented shrinkage cracks; (k) tidally cross-stratified sandy oolites, stromatolitic bioherms and biostromes; (l) oolitic/pisolitic shoal facies; (m) cemented shrinkage cracks in siltstone; (n) repeated graded storm units including intertidal mat clasts emplaced on subtidal mudrocks; (o) intertidal deformed and variably silicified stromatolitic mats; (p) cemented oolite cut by fibrous cement-lined fractures, calcified stalked cyanobacterial colony to right; (q) interbedded desiccated dolostone beds and cross-stratified oolitic sandstones; (r) siltstones with thin storm beds. Although similar facies are found in numerous formations world-wide, specific inspiration was from pre-glacial carbonates of West Africa (i), East Greenland (a–c, j–l), NE Spitsbergen (j–r), Scotland (k–l), between-glacial carbonates of NE Spitsbergen (p–r), post-glacial carbonates of NE Spitsbergen, Greenland and Scotland (d–h, q).

facies generally lack dolomite unless dispersed off-shore by storms (some ferroan dolomite also occurs in organic-rich sediments induced by bacterial breakdown of organic matter). Textures in the shallow-water dolostones are so well preserved that they can easily be mistaken for limestones in thin section. However ooid ultrastructures, and trace element contents favour a dominantly replacive origin for the dolomite. The apparent ease of dolomitization coupled with the prevalence of primary metastable carbonates point to an oceanic Mg : Ca ratio at least as high as today.

Palaeoclimates

The climatic signal in shallow marine Phanerozoic carbonates can be extrapolated from modern occurrences of: (i) skeletal assemblages, (ii) mixed skeletal–oolitic–peloidal–grapestone-bearing facies (absent at higher latitudes), and (iii) mineralogical indicators of aridity or active evaporation. In the Proterozoic oceans, devoid of skeletal organisms, the situation was necessarily somewhat different. The average supersaturation of seawater for calcium carbonate should have been higher, although undersaturation would still be expected in cooler oceans with significant freshwater input, and low evaporation. It seems likely that there was an even greater concentration of carbonate deposition in warm, actively evaporating parts of the world than today. Unlike skeletal organisms, benthic microbial mats do not precipitate carbonate from undersaturated waters today, and so stromatolitic carbonates would not be expected outside of the latitudinal range of other Proterozoic marine carbonates. Shallow-water environments would be the dominant areas for carbonate production (in the form of ooids and whitings), although planktonic photosynthesis might also allow whiting precipitation to occur offshore (Knoll & Swett, 1990).

Interpretation of the non-glacial carbonate facies as a warm-water marine assemblage is favoured by three arguments. Firstly, as discussed above, there are requirements for active evaporation. Secondly, there are occurrences of boring and benthic microfossils closely similar to modern Bahamian examples (Knoll, 1985). Thirdly, it can be inferred that primary precipitates were metastable carbonates as in modern low-latitude oceans. In contrast, low-Mg calcite would be favoured at low temperatures.

It follows that Proterozoic cool-water marine conditions should have been characterized by clastic (usually siliciclastic) sediments. These conclusions support the long-held view that thick sections of Proterozoic carbonates are evidence of warm climate (Chumakov & Elston, 1989).

Bounding carbonates

The next stage is to compare the carbonates immediately adjacent to glacial formations (particularly in successions where they are conformable) with those outlined above. Generally they belong to the same population: for most examples of a facies type next to a glacial formation, a similar example distant from a glacial formation can be found. Immediately beneath glacial units there are, for example, interbedded intraclastic limestones and limestones with spar-filled shrinkage cracks, or oolitic dolostones with fibrous cements (Fig. 1.4d), or slope facies with syn-sedimentary lithification. Above glacial units there are, for example, storm-dominated beach to shoreface intraclastic dolostones (Fig. 1.4b), or shelf (originally aragonitic) limestones. In most cases therefore the preserved sediments attest to a relatively sudden and severe climatic change.

Nevertheless, indirect evidence of the build-up and decline of ice sheets is recorded in several cases by vertical facies sequences concordant with eustatic sea-level fall prior to glaciation and rise following glaciation. Thus there is an upward transition from limestone to dolostone facies (with other evidence of shallowing) in Scotland (Fairchild, 1991) and beneath the lower glacial unit in NE Spitsbergen. A well-developed regressive hemicycle 30 m thick occurs beneath the upper glacial unit in NE Spitsbergen (Fig. 1.4d) and a shelf limestone unit locally caps slope facies as the lower glacial unit is approached in East Greenland (Fig. 1.2). Conversely, relatively deep-water limestones are found overlying certain glacial units in the western USA (Tucker, 1986) and the lower glacial unit in NE Spitsbergen, whereas the upper unit is capped by a trangressive sequence of storm-dominated clastic dolostones (Fig. 1.4b). An erosion surface at the base of this trangressive sequence reflects isostatic effects and, as is clear from studies of Holocene sea-level changes, such effects might be expected to dominate in other cases, depending on the extent of isostatic depression during glaciation.

If these inferred indications of glacioeustasy are correct, the paucity of direct evidence of climatic deterioration is remarkable. There is some possible ice-rafted and/or resedimented glacigenic sediment just a couple of metres below the onset of glacimarine sedimentation in California (Tucker, 1986), and in the southern part of central East Greenland and Scotland, but it is not certain in these cases that carbonate continued to be precipitated in the water column, rather than being reworked, after the first evidence of glacial activity.

North-east Spitsbergen provides an excellent example (Fig. 1.4d) of a terrestrial lower boundary to a glacial unit. Here a regressive sequence can be attributed to eustatic sea-level fall leading to emergence of a wide plain of early-lithified carbonates. Thus we have a classic warm-water supratidal oolitic dolostone with tepees and associated fibrous cements that is penetrated by a fracture system (Fig. 1.6) interpreted as periglacial in origin. No record of temperate conditions is recorded, but neither significant erosion nor sedimentation may have been possible in temperate conditions in this geomorphological setting. Nevertheless, an interpretation in terms of a lowering in global temperatures seems far more likely than rapid continental drift to higher latitudes.

An interesting consequence of a developing glaci-

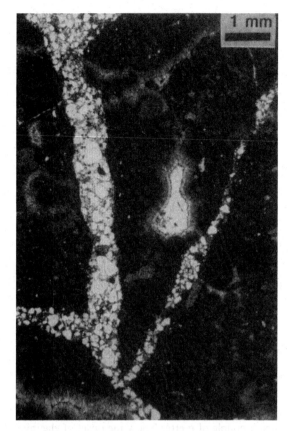

Fig. 1.6 Supratidal dolomite (as in Fig. 1.4d) with fibrous cements cut by periglacial fractures filled with quartz sand and some dolomite debris.

ation is that this may have triggered the diagenetic stabilization of underlying shelf aragonitic sediments by cooling of porewaters (in contrast it is thought that dolostones were stabilized penecontemporaneously). It is notable that the heaviest Precambrian limestone oxygen isotope values known (zero on the PDB scale) have been obtained from formerly aragonitic, slope limestones some tens of metres below the lower glacial unit of East Greenland (Herrington & Fairchild, 1989). The implication of the isotopically-heavy carbonates is that they were stabilized to calcite unusually early in diagenesis, and certainly at relatively low temperatures.

Moving to the overlying carbonates, Williams' (1979) summarized several records of 'lonestones', possibly ice-rafted, just above the top of glacial formations. It should also be mentioned that there are a number of cases of thin capping dolostones in

Australia, West Africa and North America, for example, which are incompletely understood. Although Williams' (1979) examples of cap dolostones include apparently typical warm-water oolitic types, and units with sedimentary structures similar to those of Fig. 1.4b, others, some associated with barite, some with complex deformation structures, are enigmatic. It is not clear whether this relates simply to lack of detailed study, intense diagenetic alteration, or a distinctive primary origin.

In summary, where relationships are most clearly understood, offshore carbonate sedimentation is simply sharply interrupted by glaciation, whereas in shallower-water settings marked sea-level changes are found concordant with those expected from the waxing and waning of a major ice sheet.

The glacial units

The question of whether there is a glacial input to a given Neoproterozoic formation has now been answered in most cases by a wealth of detailed studies, that of Spencer (1971) being an early landmark. Whilst Harland's (1964) belief in widespread glaciation has been vindicated, there are a number of sections where slope-related resedimentation processes unconnected with glaciation are involved.

One type of study has been the detailed sedimentological analysis of individual Proterozoic units which has led to the documentation of phenomena such as dropstone horizons (Fig. 1.7), rhythmites, striated clasts and pavements, characteristic glacial erosion phenomena, clast fabrics (Fig. 1.8) and facies sequences supporting glacial depositional mechanisms. Another notable development has been the increased understanding of glacimarine depositional processes and facies (Eyles et al., 1985) so that glacimarine rather than terrestrial glacial interpretations of depositional environments became feasible for sediments such as massive diamictites, and complexly interbedded sands, gravels and diamictites (e.g. Fig. 1.4a). For example the Scottish Port Askaig Formation with its huge number of individual diamictites and marine interbeds (Spencer, 1971) has been reinterpreted as a glacimarine succession within a rapidly-subsiding sedimentary basin (Eyles, 1989; Fairchild, 1991).

Diamictites contain abundant carbonate clasts, usually derived from lateral equivalents of underlying units, and this naturally implies that there should be a high matrix carbonate content, particularly bearing in mind the relative softness of calcite

Fig. 1.7 Top of member E1, NE Spitsbergen (Fig. 1.4e). Clast-rich diamictite is overlain by laminites with dropstones (e.g. centre), and immediately overlain by dolomitic turbidite with climbing ripple lamination. The height of the physical outcrop is 5 cm.

Fig. 1.8 Dolomite-rich, slightly fissile diamictite (within log of Fig. 1.4e), lower Varangian glacial unit (E1), NE Spitsbergen. Individual clasts can still be removed, making it possible to study clast fabrics.

and dolomite. This is the primary reason for the carbonate-rich nature of some glacial formations. A clastic origin by reworking of pre-glacial dolostones explains most beds that were previously thought to be evidence for warm conditions in glacial units. For example, many of Spencer's (1971) 'fine-grained' dolostones, turn out petrographically to be dolarenites. Nevertheless, precipitated carbonates also occur within glacial formations, but they are distinctly different from the non-glacial carbonates.

Terrestrial glacial processes

Facies analysis suggests that parts of many Neoproterozoic glacial units were deposited in terrestrial glacial settings. We should therefore consider the likely reactivity of carbonate in such environments. Figure 1.9 represents a first attempt to synthesize the possible range of behaviours of carbonate, using end-member humid and arid settings for the purposes of illustration.

The process of glacial crushing during transport naturally gives rise to large quantities of fine material (rock flour) so that glacial sediment has a high specific surface area. This leads to fast dissolution of carbonate on contact with fresh meltwater. For example, the major meltwater stream draining the modern Swiss Tsanfleuron glacier, which is based on limestone, is around half saturated for calcite. Thus in humid areas, depletion in matrix carbonate might be expected; this is a feature for example of some Precambrian West African lodgement tills (Deynoux, 1985). Given high rates of deposition in this setting, the vast bulk of the resulting dissolved material will be lost from the system. However, there are some potential mechanisms of redeposition. Regelation (the process of melting and refreezing of ice around obstacles) leads to crusts of carbonate on bedrock today, and may also be responsible for the Proterozoic calcite films along

inferred shear planes, and around clasts, described by Deynoux (1985). If depositional rates are sufficiently slow, organic breakdown reactions in lacustrine sediments could trigger concretion formation. This is known even from otherwise non-calcareous Quaternary glacial lakes (examples reviewed by Fairchild & Spiro, 1990), but has not yet been positively identified from Proterozoic rocks. Dolomite dissolution kinetics are substantially slower than calcite, but the differential properties of dolomitic and calcitic diamicts or diamictites have not yet been studied.

In arid settings, there are additional precipitation mechanisms which have been particularly well documented in the Dry Valley region of Antarctica where saline lakes are common (Wilson, 1981). Fresher lakes can have locally high rates of photosynthesis allowing precipitation either of whiting carbonate or calcareous stromatolites. Internal drainage, low precipitation and ephemeral, low discharge meltwater allow evaporation effects to become important. Although rates of evaporation are slow these small water bodies can gradually evolve to be saline, and mineral precipitation results. Although carbonates are subordinate to other salts in the modern lakes, this need not have been

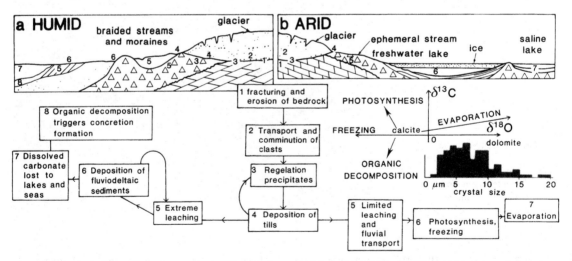

Fig. 1.9 Schematic illustration of processes in Proterozoic terrestrial glacial settings. (a) The humid cartoon is based on behaviour of modern carbonate-rich Alpine glacial systems; late Proterozoic carbonate-rich tills of Mauritania show comparable extensive leaching, and probable regelation precipitates. (b) The arid cartoon is based on the modern Dry Valleys region of Antarctica and the Proterozoic Wilsonbreen Formation of NE Spitsbergen. The arrows on the isotopic plot indicate the direction in which carbonate analyses would be shifted given the action of various processes; photosynthesis, organic decomposition, freezing and evaporation. The Wilsonbreen Formation samples lie on the line labelled evaporation. A representative histogram of crystal sizes of precipitated dolomite is also shown in Fig. 1.12.

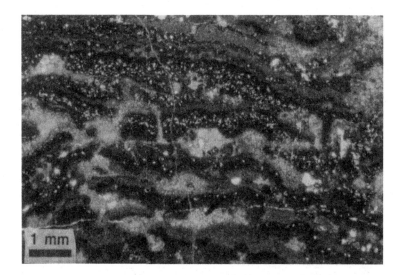

Fig. 1.10 Stromatolitic limestone with micrite layers, sediment-rich layers and syn-sedimentary fibrous cements. Interpreted as primary glaciolacustrine calcite (Wilsonbreen Formation).

the case in Proterozoic glacial systems based on carbonate.

Walter & Bauld (1983) used the Antarctican analogy to explain certain stromatolitic carbonates associated with Australian glacial deposits. However, it was not certain that the carbonates really were synchronous with glacial conditions. Well-developed columnar stromatolite inclusions between diamictites have been briefly described from Australia (Young & Gostin, 1988), but it remains to be seen whether they represent glaciolacustrine, or warm interglacial conditions.

More detailed descriptions were made of Varangian examples from Spitsbergen by Fairchild et al. (1989) which compare closely to deposits of modern saline glacial lakes. A relative retreat phase in the glaciation was marked by the occurrence of bedded sandstones, limestones and dolostones, with relatively few diamictites (Fig. 1.4c). These facies represented a coherent assemblage which in some instances were organized into a distinct retreat–advance facies sequence (Fig. 1.4ci). High rates of terrigenous input as reflected in coarse clastic beds or dropstones characterized the assemblage as a whole, supporting a glacial influence throughout deposition of the lacustrine facies. Isotopic analyses revealed a distinct evaporative trend (Fig. 1.9; Fairchild & Spiro, 1990). Such trends are known to be characteristic of primary lacustrine carbonates, and extreme gradients in $\delta^{18}O$ are known from Antarctic lakes. At the isotopically-depleted end of the range are stromatolitic limestones with well-preserved textures suggesting a

primary calcitic origin (Fig. 1.10). No well-developed columnar structures are developed, probably because of the stress of high sediment input. There are also gradations to graded calcite rhythmites which are strikingly similar to modern Antarctic examples. Developing as nodules and beds within clastic sediments are zones of precipitated dolomite (Fig. 1.11) which prove to be isotopically heavy, although the heaviest $\delta^{18}O$ values of around + 10‰ PDB were found in dolomitic stromatolites. Formation from evaporated, isotopically-heavy waters was implied. This is consistent with evidence from pseudomorphs and collapse structures, of the former presence of soluble evaporites. All the authigenic carbonates were significantly more Mn-rich than the detrital carbonates, pointing to the importance of mildly reducing conditions during carbonate formation, either in the water column, or close to the sediment surface.

In conclusion, the products of both humid and arid-zone diagenesis of tills containing detrital carbonate have been recognized in Proterozoic rocks, and there is abundant scope for future work along these lines in other areas.

Glacimarine processes

As mentioned earlier, subaqueous glacigenic facies, including diamictites have been widely recognized, and indeed probably dominate the sedimentary record of Proterozoic glaciations. Generally these represent glacimarine settings, although large lakes could also accumulate similar facies. Marine glacial

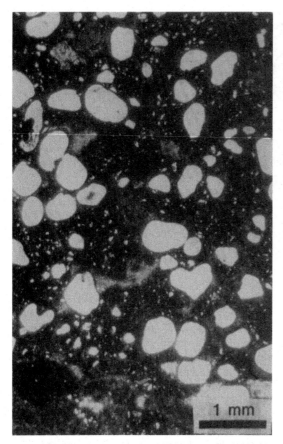

1 mm

Fig. 1.11 Sandstone in which quartz clasts are floating in a matrix of dolomite (Fig. 1.4cii). Such displacive dolomite is also characteristically rich in ^{18}O and is interpreted as a primary growth beneath a saline glacial lake.

environments may be fed by meltstreams, but a more common situation during intense glaciation is that glaciers will terminate in seawater, as illustrated in Fig. 1.12. Modern settings of this type are generally carbonate-poor, and so we have to rely more on inference and evidence from the rock record. We would expect that one potentially important difference from waterlain sediments of Fig. 1.9 is the potentially short contact time of carbonate sediment with water: thus reactive rock flour will have a greater chance of being deposited rather than completely dissolved. Also, after deposition, pore waters would be static (unlike the case for lodgement tills), so that dissolved carbonate would have to be removed by diffusion alone. These considerations would also apply to ice-contact lakes.

The carbonate matrix has been investigated in detail in an example from the dolomitic lower Varangian glacial unit in NE Spitsbergen (Fig. 1.3; Fairchild *et al.*, 1989). Field and petrographic evidence clearly demonstrated the clastic nature of the sediments (Figs 1.4e, 1.7, 1.8 and 1.13) which have structures exactly comparable to carbonate-free glacial sediments and, significantly, lack evidence of evaporative processes. Initially, study of ultra-thin sections (Fairchild, 1983) revealed that the glacial sediments had a finer modal crystal size (always less than 3 μm) than the finest of the dolostones found as clasts in diamictites (see histograms on Fig. 1.12). Such textures thus seemed to indicate preservation of a proportion of glacially-crushed rock flour, which was most prominent in quartzose lithologies (inset 6 on Fig. 1.12, and Fig. 1.14). It was recognized that the common presence of dolomite euhedra meant that recrystallization had occurred, but it was not until isotope analysis was undertaken that it became apparent that recrystallization appeared to have been syn-depositional. Compared with clasts, the matrix was depleted in ^{13}C and enriched in Fe, Mn and ^{18}O (Fig. 1.12), several samples being isotopically heavier ($\delta^{18}O$ values up to $+4.5‰$ PDB) than any other Precambrian dolomites known from outside of glacial formations. Samples with higher values of $\delta^{18}O$ were also rich in Sr and Na. These characteristics were used to argue for a preservation of the primary chemistry of the products of dolomite recrystallization under frigid marine conditions. The lightening in carbon isotopes reflected bacterial oxidation of (isotopically light) organic matter and incorporation into the recrystallizing dolomite. The spread of oxygen isotope values was taken to reflect meltwater dilution of the sea (Fairchild & Spiro, 1990), although it is also possible it could also reflect varying contamination by unrecrystallized larger grains of detrital dolomite if the total supply of light carbon was constant in each case.

The mechanism for recrystallization was hypothesized to be Ostwald ripening which has been convincingly shown to be an important mechanism of opaline silica diagenesis. Ostwald ripening is the process whereby particles of high specific surface area tend to dissolve and reprecipitate with a more compact morphology in contact with a saturated solution. Ripening is now being invoked as a pervasive mechanism in burial diagenesis. It would be slower at earth surface temperatures, but on the other hand, glacially-transported sediment should

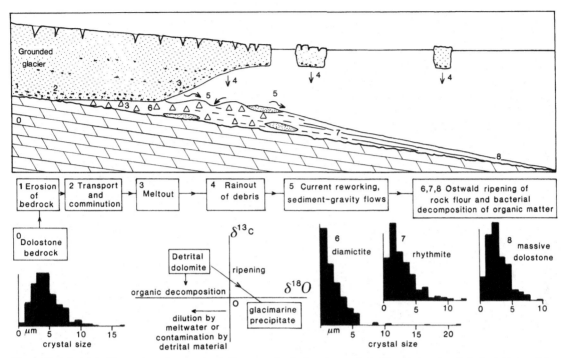

Fig. 1.12 Schematic illustration of processes in Proterozoic glacimarine settings. Based on general glacimarine models and data from the lower Varangian glacial unit at Polarisbreen, NE Spitsbergen, where the proximal (6) to distal (8) relationships are seen over a 3 km distance. Representative histograms of dolomite crystal sizes of bedrock and the matrix of glacially-transported sediments (following diagenesis) are shown together with a summary of the factors influencing the isotopic composition of the sediments (discussed in text).

be unusually reactive. Evidence for this process is now being sought in contemporary glacial settings.

In summary, the inferred process of syn-depositional recrystallization has considerable palaeoclimatic potential: if it did occur widely it could provide a method for distinguishing marine from non-marine waterlain glacial sediments.

Low-latitude glaciation

The previous discussion has highlighted the inferred great climatic shifts between glacial and non-glacial formations. The relative isotopic heaviness of the glaciolacustrine carbonates has also been contrasted with isotopically-light polar waters (Fairchild *et al.*, 1989) and can be used as evidence against a restriction of Varangian glaciation to polar latitudes. Also the prevalence of non-glacial strata and the world-wide occurrence of the glacial units tends to support a model of glacial conditions extending into low palaeolatitudes rather than warm conditions extending into high palaeolatitudes.

Much palaeomagnetic evidence of low-latitude glaciation is now available (Chumakov & Elston, 1989), but Williams' (1975) hypothesis of *preferential* low-latitude glaciation due to changes in the obliquity of the ecliptic is undermined by lack of any confirmatory evidence of exceptionally seasonal climates. Much of the evidence originally brought forward to support this view in terms of warm climates in glacial intervals (Williams, 1975) can now be discarded. The evidence reviewed in this paper points only to longer term climatic change. There was support for Williams' model in the low palaeolatitudes derived for rhythmites of the Elatina Formation of South Australia (Embleton & Williams, 1986) since the remarkable cyclicity of these deposits was interpreted as reflecting sunspot cycles, individual laminae recording seasonal melting in equatorial latitudes. This interpretation has now been replaced by an even more convincing interpretation as ebb-tidal rhythmites in a glacial setting, the cyclicity recording tidal phenomena (Williams, 1989).

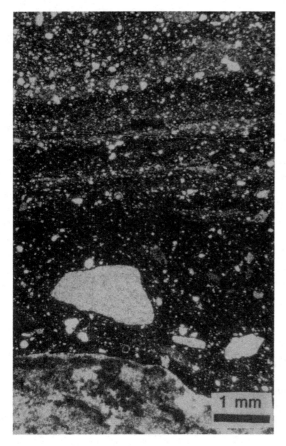

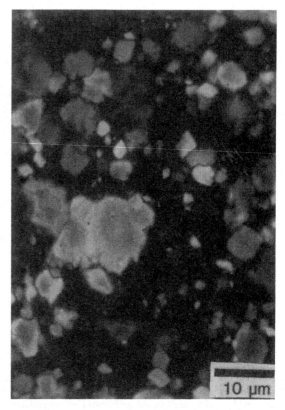

Fig. 1.14 Ultra-thin section (carbonates show first-order polarization colours) of highly dolomitic diamictite (lower glacial unit, NE Spitsbergen). Abundant sub-micron dolomite crystals still preserved (see histogram 6 in Fig. 1.12), but many crystals are euhedral. Interpreted as reflecting glacial abrasion followed by syn-depositional recrystallization by Ostwald ripening.

Fig. 1.13 Photomicrograph of laminated glacimarine diamictite (lower glacial unit, NE Spitsbergen) with conspicuous dolomite clasts. Matrix contains authigenic dolomite and pyrite.

As anticipated by Roberts (1976), thinking is now moving towards atmospheric composition being a key control of global mean temperature. It is not possible to demonstrate readily a decrease in atmospheric CO_2 from the abundance of pre-glacial carbonates as Roberts (1976) supposed, but such a decrease might be related to changing patterns of tectonic activity (Young in Chumakov & Elston, 1989 and current work). Recent climatic modelling (Hay *et al.*, 1990) does not support Young's contention that the approximately equatorial supercontinent suggested from latest palaeomagnetic work would lead to more chemical weathering and CO_2 removal. Nevertheless, as Young points out, repeated glaciation did occur during a period of incipient break-up of a supercontinent, and ceased

when this continent broke up near the beginning of the Cambrian period. The evidence is certainly for alternate states of warming and cooling, and it is notable that a tendency to shift rapidly from one stable climatic regime to another is also a feature of the Quaternary glacial–interglacial cycles.

Concluding remarks

Distinctive glacimarine and glaciolacustrine secondary carbonates have now been recognized which have considerable palaeoclimatic significance. They only exist because the glacial systems in which they formed were charged with carbonate clasts from underlying stratigraphic units which formed in warm climatic conditions. In conformable successions this warm climate can be inferred to have shifted rapidly to a glacial one and deglaciation appears to have been similarly rapid.

These conclusions reinforce Harland's (1964) notion of the usefulness of glacial units in Proterozoic chronostratigraphy. The commencement in 1991 of the International Geological Correlation Programme Project 320 (Neoproterozoic events and resources) offers the prospect of an intense period of international collaborative research in which, among other objectives, the stratigraphic framework offered by glacial units can be compared with that to become available from chemostratigraphy, biostratigraphy, sequence stratigraphy and possibly magnetostratigraphy. Only with a solid global time-stratigraphic framework can we hope to understand the underlying causes of the remarkable palaeoclimatology of Neoproterozoic times.

Acknowledgements

I thank my collaborators Michael Hambrey, Paul Herrington, Andrew Moncrieff and Baruch Spiro, and colleagues involved in IGCP Project 260 (glacial sediments through time). My work on contemporary analogues is supported by NERC Grant GR3/7687. Paul Wright and Julian Andrews made very helpful criticisms of the original manuscript.

References

Bjørlykke, K., Bue, B. & Elverhoi, A. (1978) Quaternary sediments in the northwestern part of the Barents Sea and their relation to the underlying Mesozoic bedrock. *Sedimentology* 25, 227–246.

Carey, S.W. & Ahmad, N. (1961) Glacial marine sedimentation. In: *Geology of the Arctic*, Vol. 2 (Ed. G.O. Raasch) pp. 865–894. Toronto University Press: Toronto.

Chumakov, N.M. & Elston, D.P. (1989) The paradox of Late Proterozoic glaciations at low latitudes. *Episodes* 12, 115–120.

Deynoux, M. (1985) Terrestrial or waterlain glacial diamictites? Three case studies from the Late Precambrian and Late Ordovician glacial drifts in West Africa. *Palaeogeogr. Palaeoclim. Palaeoecol.* 51, 97–141.

Embleton, B.J.J. & Williams, G.E. (1986) Low palaeolatitude of deposition for Late Precambrian periglacial varvites in South Australia: implications for palaeoclimatology. *Earth Planet. Sci. Lett.* 79, 419–430.

Eyles, C.H. (1989) Glacially- and tidally-influenced models of shallow marine sedimentation of the Late Precambrian Port Askaig Formation, Scotland. *Palaeogeogr. Palaeoclim. Palaeoecol.* 68, 1–25.

Eyles, C.H., Eyles, N. & Miall, A.D. (1985) Models of glaciomarine sedimentation and their application to the interpretation of ancient glacial sequences. *Palaeogeogr. Palaeoclim. Palaeoecol.* 51, 15–84.

Fairchild, I.J. (1983) Effects of glacial transport and neomorphism on Precambrian dolomite crystal sizes. *Nature* 304, 714–716.

Fairchild, I.J. (1991) Itineraries I, II and III. In: *The Late Precambrian Geology of the Scottish Highlands and Islands* (Ed. C.J. Lister) pp. 23–52. Geologists Association Guide 44: London.

Fairchild, I.J. & Hambrey, M.J. (1984) The Vendian of NE Spitsbergen: petrogenesis of a dolomite–tillite association. *Precambrian Res.* 26, 111–167.

Fairchild, I.J. & Spiro, B. (1990) Carbonate minerals in glacial sediments: geochemical clues to palaeoenvironment. In: *Glacimarine Environments: Processes and Sediments* (Eds J.A. Dowdeswell & J.D. Scourse). Geol. Soc. Spec. Publ. 53, 201–216.

Fairchild, I.J., Hambrey, M.J., Spiro, B. & Jefferson, T.H. (1989) Late Proterozoic glacial carbonates in NE Spitsbergen: new insights into the carbonate–tillite association. *Geol. Mag.* 126, 469–490.

Hambrey, M.J. & Harland, W.B. (eds.) (1981) *Earth's Pre-Pleistocene Glacial Record*. Cambridge University Press: Cambridge.

Harland, W.B. (1964) Evidence of Late Precambrian glaciation and its significance. In: *Problems in Palaeoclimatology* (Ed. A.E.M. Nairn) pp. 119–149, 179–184. J. Wiley: London.

Harland, W.B. & Herod, K. (1975) Glaciations through time. In: *Ice Ages: Ancient and Modern* (Eds A.E. Wright & F.M. Moseley). Geol. J. Spec. Iss. 6, 189–216. Seel House Press: Liverpool.

Hay, W.W., Barron, E.J. & Thompson, S.L. (1990) Global atmospheric circulation experiments on an earth with polar and tropical continents. *J. Geol. Soc. Lond.* 147, 749–757.

Herrington, P.M. & Fairchild, I.J. (1989) Carbonate shelf and slope facies evolution prior to Vendian glaciation, central east Greenland. In: *Caledonide Geology of Scandinavia* (Ed. R.A. Gayer) pp. 263–273. Graham & Trotman; London.

Knoll, A.H. (1985) The distribution and evolution of microbial life in the Late Proterozoic era. *Ann. Rev. Microbiol.* 39, 391–417.

Knoll, A.H. & Swett, K. (1990) Carbonate deposition during the Late Proterozoic era: an example from Spitsbergen. *Am. J. Sci.* 290-A, 104–132.

Moncrieff, A.C.M. & Hambrey, M.J. (1990) Marginal-marine glacial sedimentation in the late Precambrian succession of East Greenland. In: *Glacimarine Environments: Processes and Sediments* (Eds J.A. Dowdeswell & J.D. Scourse). Geol. Soc. Spec. Publ. 53, 387–410.

Roberts, J.D. (1976) Late Precambrian dolomites, Vendian glaciation, and synchroneity of Vendian glaciations. *J. Geol.* 84, 47–63.

Schermerhorn, L.J.G. (1974) Late Precambrian mixtites: glacial and/or non-glacial. *Am. J. Sci.* **274**, 673–824.

Spencer, A.M. (1971) *Late Pre-Cambrian Glaciation in Scotland*. Mem. Geol. Soc. Lond. 6, 100pp.

Tucker, M.E. (1986) Formerly aragonitic limestones associates with tillites in the Late Proterozoic of Death Valley, California. *J. Sedim. Petrol.* **56**, 818–830.

Walter, M.R. & Bauld, J. (1983) The association of sulphate evaporites, stromatolitic carbonates and glacial sediments: examples from the Proterozoic of Australia and the Cainozoic of Antarctica. *Precambrian Res.* **21**, 129–148.

Williams, G.E. (1975) Late Precambrian glacial climate and the Earth's obliquity. *Geol. Mag.* **112**, 441–465.

Williams, G.E. (1979) Sedimentology, stable-isotope geochemistry and palaeoenvironment of dolostones capping late Precambrian glacial sequences in Australia. *J. Geol. Soc. Aust.* **26**, 377–386.

Williams, G.E. (1989) Tidal rhythmites: geochronometers for the ancient Earth–Moon system. *Episodes* **12**, 162–171.

Wilson, A.T. (1981) A review of the geochemistry and lake physics of the Antarctic dry areas. In: *Dry Valley Drilling Project* (Ed. L.D. McGinnis) pp. 185–192. American Geophysical Union: Washington.

Young, G.M. & Gostin, V.A. (1988) Stratigraphy and sedimentology of Sturtian glacigenic deposits in the Western part of the North Flinders Basin, South Australia. *Precambrian Res.* **39**, 151–170.

2   Cretaceous climates

JANE E. FRANCIS AND L.A. FRAKES

Introduction

The Cretaceous climate has often been described as 'warm and equable', a phrase used to describe the mild uniform climates believed to characterize this period of Earth history (e.g. Frakes, 1979; Barron, 1983). The polar regions in particular were much warmer than today: there is no evidence in the rock record for the presence of ice in polar regions but instead warmth-loving floras and faunas spread into high latitudes. The equator-to-pole temperature gradient was therefore much less than it is today. One popular theory to explain these globally warm climates is that the level of atmospheric CO_2 was much higher at that time, creating a 'greenhouse' Earth.

However, more recent refined work shows that climates during the 78 m.y. of the Cretaceous were not quite so unvarying as originally thought (Sloan & Barron, 1990). New data shows distinct variations and trends of warming and cooling, and attempts have been made to quantify these variations. Much information has come from detailed oxygen isotope data from Cretaceous ocean sediments obtained during the Ocean Drilling Project (formerly the Deep Sea Drilling Project), and from the study of Cretaceous high-latitude deposits, many still situated in remote polar regions. Also important for understanding Cretaceous climates is the global distribution of fossil plant and animal assemblages and extrapolations made by comparison with climate conditions in which their nearest living relatives live. Information about past climates can be obtained from fossil plants from, for example, the shape, size and texture of leaves (Chaloner & Creber, 1990; Spicer, 1990) and from the analysis of climate patterns stored in tree-rings in fossil wood (Creber & Chaloner, 1985). To complement this, computer modelling has revealed clues about some of the limiting parameters of ancient climates.

Setting the scene

The past latitudinal position of continents on the Earth had an important influence on climate, particularly on the amount of sunlight received and on pathways of heat-carrying ocean currents. Prior to the Cretaceous all continents were assembled into the single supercontinent of Pangaea as a continuous block of land that spread from the northern to southern polar regions. The land was surrounded by the vast ocean of Panthalassa (the ancient Pacific Ocean) and by the Tethys Sea, which cut a huge wedge into the eastern margin of the continent mass along the equator. Ocean currents were able to circulate unhindered as large gyres from the equator to the poles, carrying with them warmth from the low latitudes which helped keep the polar regions free of major ice caps (Fig. 2.1) (Haq, 1984).

By the early part of the Cretaceous the continents were, however, beginning to split apart. The North American continent moved away from Eurasia, allowing the development of the Central and North Atlantic oceans. Ocean circulation began in the central Atlantic region. As the split gradually widened to meet up with the Tethys Sea, a circum-equatorial westward current was established for the first time and the southern continents (Gondwana) were then separated from those in the north (Laurasia). Later, however, the movement of small continental blocks in the Panama region blocked this circum-equatorial current, deflecting warm currents northwards to form the proto-Gulf Stream (Gradstein & Sheridan, 1983).

The South Atlantic Ocean developed later in the Early Cretaceous. At first the enclosed seas that formed in the rift sites provided suitable environments for the formation of extensive evaporite deposits. The Falkland Plateau, which extended eastward from the southern tip of South America, was high enough at first to block ocean connections with the open ocean to the south. It was only during

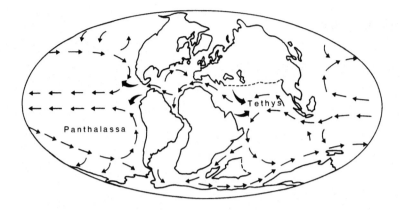

Fig. 2.1 Reconstruction of palaeogeography for the middle Cretaceous, showing continental positions and proposed pathways of ocean currents. The large arrows are potential sites of warm saline bottom-water formation. (Modified from Haq, 1984.)

the time of high sea-levels in the Aptian/Albian that circulation in the South Atlantic was finally linked to the global ocean.

By the later part of the Cretaceous, ocean circulation was still dominated by huge gyres in the Pacific that travelled from the equator into high latitudes (Haq, 1984; Barron & Peterson, 1990). The Tethys Sea was still open, although circulation patterns became disrupted when India moved north across the ocean. Circulation in the Atlantic oceans was becoming well-established, with the South Atlantic connected to the Indian Ocean between Africa and Antarctica. The rift between Australia and Antarctica was also developing into an important seaway by this stage (Veevers & Ettriem, 1988), but a complete circum-Antarctic current could not develop to isolate Antarctica due to the connection via the Scotia Arc between Antarctica and South America. The isolation of Antarctica by cold circumpolar currents, which led to the build-up of the present ice cap and the last ice age, began later in the Tertiary.

The warm oceans

Analysis of the oxygen isotopes in the shells of marine organisms has proved to be an important source of information about Cretaceous ocean temperatures. The ratio of the isotopes ^{18}O and ^{16}O preserved in the mineral skeletons of small marine organisms, such as Foraminifera and nanoplankton, is a record of the isotopic composition of the seawater during the life of the organism. This ratio was originally influenced by the water temperature and/or the presence of ice (ice contains more ^{16}O, leaving the seawater relatively enriched in ^{18}O).

Temperature estimates of Cretaceous oceans

have been obtained from microfossils from the northwest Pacific (Douglas & Woodruff, 1981) (Fig. 2.2). These data from planktonic and benthic Foraminifera indicate that at the beginning of the Cretaceous low-latitude sea-surface temperatures were about 28°C and bottom waters about 17°C, giving a depth gradient of about 11°C. Ocean temperatures cooled slightly from the Berriasian to the Barremian by 2 or 3°C and then rose to a peak during the late Albian/early Cenomanian. Temperatures dropped slightly from the Cenomanian to the Santonian, followed by a warming in the Campanian and a final cooling episode during the Maastrichtian.

The planktonic and benthic isotope curves of Douglas & Woodruff (1981) indicate that, even though both surface and deep ocean temperatures varied over time, they changed in unison and so the vertical temperature gradient remained stable. This implies that the ocean was reacting to temperature changes as a uniform mass, compared to the present ocean which has a distinct deep-water layer formed of cold-water sinking down from the surface in the polar regions.

Temperatures for equatorial oceans have been obtained from oxygen isotopes in phosphatic remains of fish in Israel (10°KrN) (Kolodny & Raab, 1988). The data show that in the early Cretaceous shallow ocean waters were nearly 30°C, peaking at 33°C in the early Turonian, then cooling to a low of about 20°C in the late Campanian to the early Maastrichtian with a final warming trend into the Palaeocene. This is not disimilar from the trends in the equatorial Pacific. There is, however, a reasonably consistent temperature difference of about 10°C between high and low latitudes (Fig. 2.2), suggesting that warming and cooling episodes during the Cretaceous were global in extent.

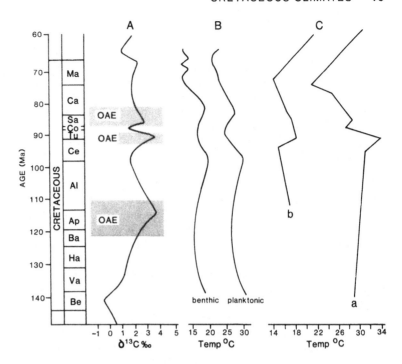

Fig. 2.2 Compilation of isotope data for the Cretaceous. (A) The average carbon isotope composition of C-org free pelagic limestones; the shaded areas represent times of ocean anoxic events (from Arthur *et al.*, 1990). (B) The oxygen isotope temperature curves from planktonic and benthic Foraminifera, NW Pacific (summarized from Douglas & Woodruff, 1981). (C) Temperature curves from oxygen isotopes of phosphatic remains of fish from Cretaceous equatorial waters in Israel (a) and a summary (through the most O-rich points) for high-latitude carbonates (b) (from Kolodny & Raab, 1988).

Information about the temperatures of high-latitude oceans is sparse but some temperatures have been obtained from isotope measurements in belemnites, summarized in Stevens (1981). In general they show a trend of cooling in the early Cretaceous, a mid-Cretaceous peak of warmth followed by further cooling. Isotopes from benthic Foraminifera from about 65°KrS near the Antarctic Peninsula show that a significant cooling phase occurred during the late Maastrichtian. Shelf temperatures cooled from about 5.5–9°C during the mid-Maastrichtian to 4–8.5°C during the late Maastrichtian, according to Barrera *et al.* (1987), or from 13.6°C in the Santonian/Campanian to 11.7°C in the Maastrichtian, according to Pirrie & Marshall (1990).

Although there are some limitations to the reliability of isotope analyses (Barron, 1983; Hudson & Anderson, 1989), these temperatures indicate that equatorial ocean temperatures were similar to today but high-latitude seas were probably much warmer. The equator-to-pole temperature gradient in the oceans was much less during the Cretaceous than at present: Barron (1983) suggested a pole-to-equator temperature contrast of between 17°C and 26°C for the Cretaceous, compared to 41°C for the present. From this it has been implied that circulation was sluggish and the waters poorly mixed. These warm oceans would have also been rather stagnant and had a low oxygen content. The lack of polar ice caps implies that there was no major source of cold oxygen-rich bottom water, as there is today; instead Cretaceous deep-ocean waters may have consisted of a mix of waters from cool high latitudes and from highly saline surface water from low latitudes. Bottom waters as warm as 22°C during the Coniacian in the Angola Basin (South Atlantic) (Saltzman & Barron, 1982) have promoted the idea that in some places in mid- to low latitudes the deep-ocean waters may have been formed of warm, dense and highly saline waters, which formed in marginal seas by evaporation and then sunk into deeper levels of the ocean (Brass *et al.*, 1982). However, Barrera *et al.* (1987) concluded that the cold-water forming on the shelves around Antarctica was a source of cold bottom water for the Pacific during at least the later part of the Cretaceous.

Climates on land

Fossil plants have provided one of the main sources of land climate information for the Cretaceous. In the past, the climate has been deduced from comparison with the thermal tolerances of living relatives and the latitudinal distribution of the fossil zones (Chaloner & Creber, 1990). Although useful as a general guide this is now proving to be a rather suspect method because the climatic tolerances of some ancient plants may have been broader than they are today. A more refined analytical approach can be used on Late Cretaceous angiosperm plants, that of physiognomic leaf analysis, using specific leaf characteristics such as shape and size, which can be correlated with climate.

Fossil floras from Lower Cretaceous rocks appear to be composed of a fairly ubiquitous collection of ferns, conifers and cycadophytes, which grew throughout low to high latitudes. Their division into floral provinces was broadly correlated with tropical/sub-tropical climates in the equatorial region and temperate climates in higher latitudes (Vakhrameev, 1975; Hallam, 1985). In the northern hemisphere the vegetation was dominated by the Cheirolepidiacean conifers, an extinct family of conifers that shed *Classopollis* pollen. Since pollen of this type has frequently been found in association with evaporitic sediments its abundance has been used as an indicator of arid climates, particularly for Jurassic and early Cretaceous times in the USSR (Vakhrameev, 1981). On this basis, peak aridity occurred during the late Jurassic but in the early Cretaceous conditions appear to have become much more humid as the abundance of *Classopollis* declined and ferns became more dominant. *Classopollis* pollen was rare in sediments in the Aptian/Albian and late Santonian/early Campanian, implying humid climates and, more common during the

Turonian and the early Maastrichtian, times of relative aridity.

One of the most intriguing aspects of the Cretaceous world was that luxurious forests, inhabited by dinosaurs, grew well up into the polar regions. For example, rich fossil floras have been described from Alaska (Parrish & Spicer, 1988), the Siberian region (Krassilov, 1981), Australia (Douglas & Williams, 1982; Jell & Roberts, 1986), and Antarctica (Dettman, 1989), all sites that were in high latitudes during the Cretaceous. In Antarctica forests of podocarp and araucarian conifers grew up to 70°S on the Peninsula. Features of the growth rings in the fossil wood from these forests are comparable to those in living trees in cool-warm temperate forests in Australasia (Jefferson, 1982; Francis, 1986) (Fig. 2.3).

It was previously considered that it was impossible for normal temperate vegetation to survive at such high latitudes and that the plants would not have been able to endure the long periods of winter darkness. When palaeomagnetic studies confirmed that the continents had indeed been situated near the poles in the Cretaceous it was proposed that there must have been a decrease in the axis of tilt of the Earth. This would have allowed sunlight to reach the polar lands even during the winter. However, apart from the fact that no realistic mechanism could be found that would have radically changed the Earth's tilt, modelling by computer showed that, with a decrease in the Earth's tilt, sunlight reaching high latitudes would have been of such low intensity that it would not have been

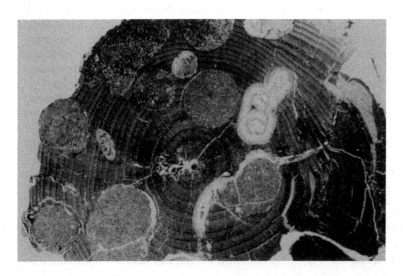

Fig. 2.3 Petrified conifer wood from Upper Cretaceous sediments on the Antarctic Peninsula, from about 65°KrS. The growth rings indicate that the climate was favourable for tree growth and very seasonal. The circular areas are the sediment-filled borings of the marine borer *Teredolites*. Specimen diameter is 6 cm.

sufficient for plant growth (see discussion in Barron, 1984). It is now considered that the forests were well-adapted to polar climates, probably becoming dormant during the dark winter months, by being deciduous, and by having conical shapes to avoid mutual shading in the low-angled light (Creber & Chaloner, 1985).

Dinosaur bones are also present in Cretaceous polar deposits. Whether these supposed warmth-loving animals were able to survive the months of winter darkness, especially if it was cold, has provoked much debate. Some propose that they were able to cope with such conditions and stayed there all year round; others consider that they would have had to escape the cold and dark by migrating to warmer lower latitudes (Parrish *et al.*, 1987; Rich & Rich, 1989).

One of the criteria used to assess the Cretaceous climate in the past has been the presence of cycadophytes, a common component of Mesozoic fossil plant assemblages. It was assumed that they were tropical, warmth-loving plants, like many of the living cycads. However, it has been shown that some fossil cycadophytes have a different morphology and were probably deciduous. Since they were common in high-latitude floras, such as those in Alaska (Spicer & Parrish, 1986), they were probably well-adapted to growing in somewhat cooler seasonal climates.

The mid-Cretaceous saw a great change in vegetation with the appearance of the flowering plants. Analyses of leaf characteristics of angiosperm fossil assemblages have provided a great deal of detailed information about Late Cretaceous climates, particularly in North America. Based on this technique, Wolfe & Upchurch (1987) have proposed mean annual temperatures of approximately 21°C in the Albian at a palaeolatitude of about 30°N, rising to a warm peak of 25°C during the Santonian. From then temperatures cooled to 23–24°C during the Campanian and 22°C in the early Maastrichtian. At higher latitudes in Alaska similar leaf and vegetation analyses by Parrish & Spicer (1988) indicated that the mean annual temperature during the latest Albian and Cenomanian was about 10 ± 3°C, slightly warmer in the Coniacian (12–13°C), followed by a cooling during the Campanian and Maastrichtian to temperatures of about 2–8°C (Fig. 2.4). Distortion of leaf forms and the abundance of coals formed from undecayed leaf litter suggested to Spicer (1987) that winter temperatures may have declined to freezing or below. Spicer & Parrish

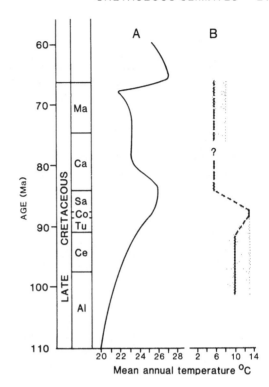

Fig. 2.4 Estimated temperatures for the late Cretaceous from land plants. (A) Leaf margin analysis of plants from low to mid-palaeolatitudes of North America (from Wolfe & Upchurch, 1987). (B) Analysis of plants from the North Slope of Alaska (from Parrish & Spicer, 1988). In (B) the solid line represents estimated mean annual temperatures (MATs) from leaf margin analysis, the heavy dashed line represents the MATs estimated from vegetational structure, and the shaded area represents the range of MAT.

(1990) suggested that periglacial conditions did not develop at sea-level during the Maastrichtian but permanent ice was likely above 1000 m at 85°N. These two sets of data indicate that the temperature gradient during the late Cretaceous was about 0.3°C per degree of latitude, less than that at present (about 0.5°C per °lat).

Climate-sensitive rocks

The distribution of some rock types whose formation was related to aspects of climate, particularly rainfall and evaporation, hold clues to ancient climate changes. The distribution of coals, evaporites, bauxites and other indicators for the Cretaceous have been recorded in detail by Parrish *et al.*

(1982) and Hallam (1984). The distribution of rainfall was also predicted by Parrish & Curtis (1982) using qualitative models, based on the relation of modern rainfall patterns with Cretaceous geography and topography. The models predicted regions of relative aridity in low to mid-latitudes on the western sides of continents and in the large landmass of Asia. High-latitude regions and the margins of Tethys had high rainfall (Fig. 2.5). The size of the arid zones decreased during the Cretaceous, reflecting increasing global humidity. Hallam (1985) related this to increasing maritime influence as continents were becoming more dispersed and sea-levels higher.

There was a significant increase in the amount and distribution of evaporites during the latest Jurassic from the preceding times. Evaporites formed in latitudes of up to 40°N and 35°S, with coals in high latitudes. However, during the Early Cretaceous the amount of evaporites decreased markedly and were mainly concentrated in low latitudes, indicating that the climate had become more humid. Frakes (1979) showed some evaporites at higher latitudes, but these may have been related to the high sea-levels during the Cretaceous, which created a large expanse of marginal marine environments particularly suited for the formation of evaporites. There were also several enclosed basins in existence at this time that formed as continents moved apart, such as those at the rift site of the South Atlantic. Restricted circulation in these would have been favourable for evaporite formation.

The change to more humid climates during the Cretaceous is also reflected in the change in clay mineral suites in sediments. For example, in north-west Europe in the late Jurassic to the Valanginian,

kaolinite increased in abundance and mixed-layer minerals decreased, reflecting a change from hot, semi-arid climates to warm humid-temperate climates (Sladen, 1983). Soils changed from alkaline and poorly-leached to acid and well-leached podsols. During this interval the land areas in Europe changed from low relief to relatively high land, which may have influenced source area climate to some extent, and the expanding North Atlantic may have contributed to the development of more humid climates. However, from the Hauterivian to the Aptian mixed-layer clays increased in abundance at the expense of kaolinite, reflecting lowered source area relief and perhaps climate warming. A change from carbonate to clastic deposition in the area (Hallam, 1985), and a change in the patterns of growth rings in wood that grew in the region (Francis, 1987) also reflects the climate change. This variation in clay minerals has also been recognized in many other localities world-wide (Hallam, 1985).

By the Late Cretaceous the climate had continued to be humid, as indicated by the scarcity of evaporites. Coal and ironstones occurred in north-west Africa and northern South America as humid conditions developed in the equatorial region for the first time. Coals were still abundant in high-latitudes, where rainfall was moderately high (Parrish & Curtis, 1982; Hallam, 1984).

The distribution of reefs is an important clue to climate in the past. However, in the Cretaceous, reefs were widespread to about 30° latitude, more than in previous times, and particularly abundant along continental margins bordering the new equatorial ocean passage. Many of the reefs were composed of rudists, bivalves resembling corals in shape but perhaps with different ecological tolerances.

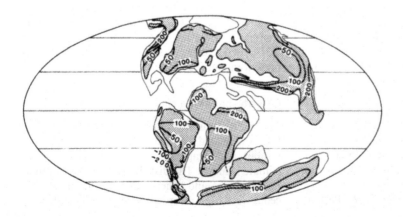

Fig. 2.5 Predicted distribution of rainfall for the middle Cretaceous. The dark shaded areas represent land and the lightly shaded areas represent the edge of the continental shelf. The rainfall is categorized as follows: <50 low rainfall, 50–100 moderately low rainfall, 100–200 moderately high rainfall, >200 high rainfall. (Modified from Parrish et al., 1982.)

Carbonates spread up to 30° north and south, similar to their spread in former times: Ziegler *et al.* (1984) thus concluded that the limiting factor was not temperature, but light availability.

By comparison with their results of predicted rainfall Parrish & Curtis (1982) concluded that the distribution of coals and evaporites in the Cretaceous was best explained by a zonal model of climate, that is, controlled by latitudinal temperature changes. Similar conclusions had been made for the distribution of floral assemblages, the floral boundaries drawn approximately parallel to latitudes (Vakhrameev, 1975). However, the results of climate modelling of Barron & Washington (1985) and others, discussed later, imply that there was more of a continental effect, the temperature changes occurring from the centres of large landmasses towards the margins, with seasonal extremes in continental interiors. This conflicts with the presence of some floras and faunas in mid-latitude, mid-continental positions that may not have been able to survive freezing conditions, according to present interpretations of their thermal tolerances (Archibald, 1991; Wing, 1991). More integrated studies by climate modellers and palaeontologists are needed to resolve this problem (Sloan & Barron, 1991; Wing, 1991).

The carbon record

During the Cretaceous vast quantities of organic-rich black shales were deposited. Although not directly related to climate, these carbonaceous deposits were an integral part of the complex carbon cycle which includes interaction with oceanic conditions, atmospheric CO_2, and thus climate.

Most black shales were deposited in several discrete 'events' during the mid- to late Cretaceous: there was a long phase of deposition during the Aptian/Albian (117–100 Ma), a brief but worldwide event in the Cenomanian/Turonian (93–90 Ma) and a minor episode during the Santonian/Campanian. Some black shales were also deposited during the Valanginian to the Barremian. Deposition took place in many situations, including oceans, continental basins, on submarine highs, and in shallow-shelf environments. These episodes of black-shale deposition have been correlated with ocean anoxic events, high sea-levels and positive excursions of the $\delta^{13}C$ carbon isotope record (Arthur *et al.*, 1990) (Fig. 2.2.).

There is, however, still some debate about the most influential factors involved with the deposition of these organic-rich sediments (see Hay, 1988, and discussion in Arthur *et al.*, 1990). Some propose that the organic carbon (^{12}C) was preserved because the deep waters of the Cretaceous oceans were warm and stagnant and had a low oxygen content. Organic carbon that settled into these anoxic waters would not have been oxidized but would have been preserved in basinal sediments.

In contrast, others suggest that anoxia was not a necessary prerequisite, but that the rate of productivity was most important, either in the oceans or of terrestrial plant matter (for example, see Pedersen & Calvert, 1990). Marine organic carbon is primarily produced at the surface of the ocean by phytoplanktonic organisms, requiring both light and an ample nutrient supply for photosynthesis (Arthur, 1982). During this process CO_2 from the atmosphere is taken up and the carbon fractionated into the isotopes ^{13}C and organic carbon ^{12}C. The supply of nutrients (particularly phosphorus) to the surface organisms for this process is crucial: it depends on either more nutrients being washed in from adjacent land or nutrients being recycled from decayed organic matter and returned to the ocean surface by upwelling currents. Thus when the oceans are actively mixed and nutrient supply is good the productivity of organic carbon is higher, and presumably so high that a great deal of it is preserved regardless of the oxygen levels in the deep ocean. The opening of ocean gateways as a result of rifting and the subsequent mixing of water masses may have enhanced the supply of nutrients for increased productivity, or sea-level rise may have also increased productivity and expanded the oxygen-minimum zone in the ocean (Hallam, 1987). All of these processes may have been operating at times during the Cretaceous (see review in Arthur *et al.*, 1990).

Periods of black-shale deposition during the Cretaceous have also been correlated with positive excursions of $\delta^{13}C$ (Fig. 2.2). In many parts of the geological record such peaks in the carbon curve correlate with intervals of cooler climates (Frakes *et al.*, in press). The formation of black shales in the Cretaceous is usually considered to have been related to the warm climate and the consequent stagnant oceans: some black shales were deposited during peaks of ocean warmth (see Fig. 2.2), but not all—some were deposited during the Valanginian to the Barremian in times of cooler climates, and Arthur *et al.* (1988) suggested that a phase of cool climate was associated with organic carbon burial at the Cenomanian/Turonian boundary. The carbon cycle is very complex but understanding the processes involved in it are crucial to understanding climate change now and in the past.

Evidence of ice

The presence of faunas and floras living in polar regions and the lack of reported glacial tillites has led to the Cretaceous generally being considered as an 'ice-free' period. However, there have been several reports of deposits formed by ice transport in early Cretaceous high-latitude sites (Fig. 2.6) (see Frakes & Francis, 1988).

The deposits are all fine-grained marine mudrocks and siltstones that contain outsized clasts of exotic lithologies. The fine-grained matrices are often laminated, indicative of quiet-water deposition below wave base, rather than having been emplaced by current activity that moved sediments laterally. The exotic clasts, sometimes up to a few metres in diameter, occur as lonestones or in pod-

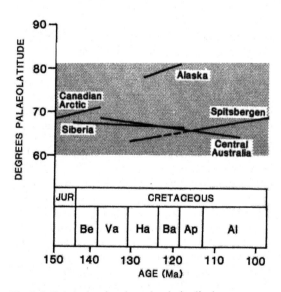

Fig. 2.6 Graph showing the palaeolatitudinal distribution of reported Lower Cretaceous boulder shales versus age. The boulder shales or ice-rafted deposits are spread between about 60° and 80° palaeolatitude, indicating that this was the zone of winter ice formation. (From Frakes & Francis, 1988). Jur, Jurassic; Be, Berriasian; Va, Valanginian; Ha, Hauterivian; Ba, Barremian; Ap, Aptian; Al, Albian.

like groups. Their close resemblance to ice-rafted dropstones in known glacial deposits suggests that they have been emplaced in the sediments by being dropped from rafts, the most likely raft being floating ice.

In northern high latitudes, shales containing outsized boulders, some of which are considered to have been ice-rafted, have been reported from late Jurassic and early Cretaceous deposits along the arctic borderlands of Siberia (collated in Hambrey & Harland, 1981). At this time northern Siberia was situated in very high latitudes, and being part of a large continental landmass would have experienced extremes of seasonal temperatures. To the west, on Spitsbergen and on the Canadian Arctic islands, early Cretaceous dark shales contain exotic rounded clasts of Palaeozoic cherts, considered to have been emplaced by ice.

In the southern hemisphere early Cretaceous ice-rafted deposits have been reported from central Australia, an area which was situated at above 65°S at that time. Large boulders of Proterozoic rocks occur within Cretaceous mudstones within the Eromanga Basin, a large epicontinental sea that covered central Australia during the Cretaceous (Frakes & Francis, 1988). Ice may have also been present further to the south in Australia, in valleys formed by the rifting of Australia and Antarctica (near 80°KrS). Calcite concretions in these continental fluvial sediments have yielded $\delta^{18}O$ oxygen isotope values near – 20‰, the depletion in ^{18}O interpreted by Gregory *et al.* (1989) as an indication that the concretions formed from glacial meltwaters. Mean annual temperatures of less than 5°C and perhaps even as low as – 6°C have been proposed by Rich & Rich (1989), though the climate was probably seasonally, rather than uniformly, cold.

All these high-latitude boulder shales also contain fossil wood and glendonite nodules. Growth rings in the fossil wood (conifers) are conspicuously narrow and contain a record of markedly seasonal growth, probably the result of low temperatures (Frakes & Francis, 1990). The occurrence of glendonite nodules within the boulder shales has also been considered further evidence of cold climates. Glendonite nodules are found as stellate aggregates of carbonate crystals (Fig. 2.7). However, these have pseudomorphed the original mineral, most commonly a calcium carbonate hexahydrate called ikaite, ·which is stable only in very low temperatures. Glendonites were first reported from Permian

Fig. 2.7 Glendonite nodules from the Lower Cretaceous Bulldog Shale in the Eromanga Basin of central Australia (see Frakes & Francis, 1988). Scale bar represents 1 cm.

glacial deposits in Australia and have since been found in most Cretaceous boulder shales (Kemper, 1987; Frakes & Francis, 1988), implying that marine conditions were very cold.

The ice-rafted deposits have, so far, been reported only from early Cretaceous sequences, suggesting that the late Cretaceous polar climates may have been warmer (except for the late Maastrichtian when there may have been montane ice and freezing temperatures in winter, Spicer & Parrish, 1990). The fossils of temperate faunas and floras also occur in these deposits and suggest that the climate was only seasonally cold to allow the formation of ice on rivers and shorelines during the cold winters, but with summers warm enough for the survival of warmth-loving biotas. Kemper (1987) further proposed that the Cretaceous climate was cyclical, with cold periods lasting for up to 2 m.y. separated by phases of warmer climates, possibly related to orbital cycles.

Figure 2.6 illustrates that seasonal ice formed in latitudes of about 60–80°. Permanent glaciers may have existed at high altitudes and higher latitudes, although direct evidence for these is lacking at present, possibly due to the erosive nature of glacial deposition (Frakes & Francis, 1990) and the problem that any signs of polar continental glaciation (in Antarctica) are covered with the present ice sheet.

Climate modelling

Some insight into Cretaceous climates has also been provided by numerical computer models. Experiments to model a warm, equable Cretaceous climate have never really been able to produce the globally warm conditions reported (Crowley & North, 1991). The models incorporated Cretaceous palaeogeography and models of present atmospheric circulation, and were prescribed to have warm ocean waters at high latitudes (10° or 20°C), as indicated by oxygen isotope data (Barron & Washington, 1982; Schneider et al., 1985; Sloan & Barron, 1990). Other models have investigated the effect of the size of continental landmasses and their distribution upon global climates (Crowley et al., 1986; Kutzbach et al., 1990).

The models predicted that, even with warm polar oceans and a reduced pole-to-equator temperature gradient, the temperatures of continental interiors could not have remained warm in winter. Large continental landmasses have low heat capacity (Crowley & North, 1991) and so the heat acquired during the Cretaceous summers would not have been retained in winter, particularly at high latitudes where the winter months were dark. The weak atmospheric circulation was not strong enough to transport warmth from the ocean into continental interiors and so they experienced low temperatures during the winter, often well below freezing (Fig. 2.8). Similarly, in summer continental interiors were heated to high temperatures. The climate of these interior regions was therefore very seasonal. For example, the modelling results of Barron & Washington (1982) produced summer temperatures of about 17°C and winter temperatures of – 23°C for Siberia, and 17°C summer and – 13°C winter temperatures in central Australia during the mid-Cretaceous. Both modelling results and recent field-work indicate that the climate was, in fact, quite seasonal with cold winters but hot summers, particularly in continental interiors. The maritime influence on regions adjacent to coasts would probably have allowed more uniform climates on continental margins.

The energy-balance models of Crowley *et al.* (1987) also demonstrated that large continental interiors would have had large seasonal temperature changes. They suggested that in order to maintain an ice-free Earth (one without permanent ice) it

was necessary to have high summer temperatures to melt winter ice. They proposed two states for non-glacial climates like the Cretaceous: 'cool non-glacials' when winters were cold but permanent ice cover was prevented by warm summers, and 'warm non-glacials' with higher average temperatures. Geological evidence suggests that the early Cretaceous, with its winter ice, might be described as 'cool non-glacial', and the late Cretaceous as 'warm non-glacial' since there is no evidence for winter ice at high latitudes (apart from during the Maastrichtian perhaps).

Apart from annual seasonality, it has been proposed that there were periods or cycles of warmer and cooler climates during the Cretaceous (Epshteyn, 1978; Kemper, 1987). Kemper proposed alternating cycles of warm and cool climates of about 2 m.y. duration, possibly caused by variations in the Earth's orbital motions, with even longer cycles of 20 m.y. superimposed on this. Cycles of Cretaceous limestones and marls described from several localities have also been attri-

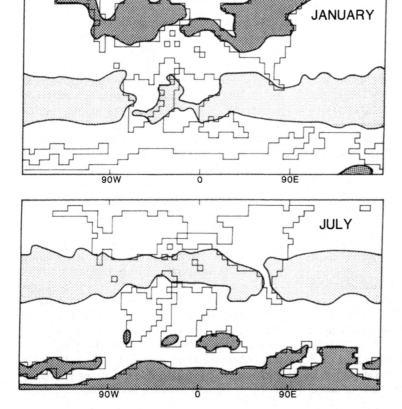

Fig. 2.8 Results of climate modelling for the middle Cretaceous. The figure shows the predicted mean temperatures at ground level for January and July in an experiment where the minimum sea surface temperature was set at 10°C. The dark shaded area represents regions with temperatures below 0°C (273 K) and the lightly shaded area represents temperatures above 27°C (300 K). (Modified from Barron & Washington, 1982.)

buted to cyclical changes in climate related to Milankovitch-type variations in orbital parameters (particularly the tilt of the orbital axis), which during the Pleistocene affected solar insolation and ice volume (reviewed by Fisher *et al.*, 1990).

Limestone–marl sequences have been attributed to cycles of carbon productivity and burial (the organic-rich marl having been formed during periods of deep water anoxia) or to changes in precipitation. For example, Barron *et al.* (1985) proposed that limestone–marl alternations in the Cenomanian/Turonian Greenhorn Cyclothem in western North America were caused by changes in climate, the limestone being deposited in times of more arid climates and the marls related to periods of greater humidity and runoff. They suggested that changes in orbital parameters in the warmer Cretaceous influenced the intensity of seasonality, land–sea thermal contrasts, monsoonal circulation and the intensity of rainfall. Similar variations discovered in oxygen isotope data from rhythmically-bedded chalk–marl sequences in southeastern England were interpreted by Ditchfield & Marshall (1989) as climate alternations between warm temperatures when chalk was formed and cooler temperatures during deposition of the marl.

It has been proposed that the Cretaceous had a 'greenhouse' climate with high levels of CO_2 in the atmosphere in order to produce the global warmth and above-freezing temperatures interpreted from faunas and floras, and particularly to warm up the polar regions. When the first climate models predicted that high-latitude continental interiors could not be warmed simply by having warm polar oceans, it was proposed that external forcing with higher levels of CO_2 would have been required (Berner *et al.*, 1983; Barron & Washington, 1985). Levels of CO_2 several times those at present have been proposed, but now the discovery that freezing temperatures did occur on the Cretaceous Earth, at least in the early Cretaceous, verify the results of the climate models and suggest that CO_2 levels need not have been so high. The late Cretaceous appears to have been warmer and CO_2 levels possibly higher, a source for which may have increased output of CO_2 from volcanic activity. Arthur *et al.* (1985) reported that during the late Cretaceous there was an increase in rifting and mid-plate volcanism, particularly in the Pacific Ocean region, which would have increased levels of CO_2 in the atmosphere (although this may have been compensated to some extent by the increased burial of organic carbon at this time).

Summary

As all the evidence of Cretaceous palaeoclimates is drawn together its appears that, far from being globally warm throughout the 78 m.y. of the Cretaceous, this was a time of quite variable climates. Although there may be some complications with the interpretation of absolute temperatures with some of the palaeoclimate data (see Parrish, 1987), the similarity in trends of temperature change is quite consistent from several sources.

At the beginning of the Cretaceous, climates were changing from an episode of global aridity at the end of the Jurassic to much more humid environments through the Cretaceous. Ocean isotope data indicate a cooling phase in the early Cretaceous: this is the time when the polar regions were seasonally cold enough for the formation of ice, and ice-rafted deposits with glendonites were formed. Evidence from tree-rings and sediments suggests that the climate was strongly seasonal at this time, with marked differences between winter and summer conditions.

There are no records of ice-rafted deposits from the late Albian onwards, indicating that high-latitude climates became much warmer. However, the interval from the Cenomanian to the Santonian appears to have been a time of quite variable climates as there is conflicting evidence for both warming and cooling. Isotope data from the Pacific indicates that ocean waters were cooling from the Cenomanian to the Coniacian, though warming again during the Santonian, whereas those from equatorial Israel seem to place the peak of warmth in the early Turonian. On land Cretaceous floras in mid-North America indicate a peak of warmth during the Santonian, while in Alaska the warmest period appears to have been during the Coniacian. To some extent this confusion may be due to correlations using different time scales. Also the dating of much Cretaceous strata is not as refined as the palaeotemperature data obtained from them. Before much of this climate information can be used to its full extent more work needs to be undertaken on detailed dating of Cretaceous sequences.

However, it is quite clear that during the final part of the Cretaceous there was a distinct cooling trend. Both ocean and land temperatures dropped considerably. The latest part of the Maastrichtian appears to have been especially cool, with low temperatures, even freezing, at high latitudes and conditions perhaps similar to those of the early Cretaceous. All climate evidence points to a warming again during the Palaeocene. Perhaps the 'catastrophe' that hit the biota at the end of the Cretaceous was due to the shock of cold temperatures after the mid-Cretaceous warmth?

References

Archibald, J. D. (1991) Comments and reply on "Equable' climates during Earth history?' *Geology* **19**, 539.

Arthur, M.A. (1982) The carbon cycle–controls on atmospheric CO_2 and climate in the geological past. In: *Climate in Earth History* (Eds W.H. Berger & J.C. Crowell) pp. 55–67. National Academy Press: Washington.

Arthur, M.A., Dean, W.E. & Schlanger, S.O. (1985) Variations in the global carbon cycle during the Cretaceous related to climate, volcanism, and changes in atmospheric CO_2. In: *The Carbon Cycle and Atmospheric CO_2: Natural Variations Archean to Present* (Eds E.T. Sundquist & W.S. Broecker) pp. 504–529. American Geophysical Union: Washington.

Arthur, M.A., Dean, W.E. & Pratt, L.M. (1988) Geochemical and climatic effects of increased marine organic carbon burial at the Cenomanian/Turonian boundary. *Nature* **335**, 714–717.

Arthur, M.A., Jenkyns, H.C., Brumsack, H.J. & Schlanger, S.O. (1990) Stratigraphy, geochemistry, and palaeoceanography of organic carbon-rich Cretaceous sequences. In: *Cretaceous Resources, Events and Rhythms* (Eds R.N. Ginsburg & B. Beaudoin) pp. 75–119. Kluwer Academic Publishers: The Netherlands.

Barrera, E., Huber, B.T., Savin, S.M. & Webb, P.N. (1987) Antarctic marine temperatures: Late Campanian through Early Eocene. *Paleoceanogr.* **2**, 21–47.

Barron, E.J. (1983) A warm, equable Cretaceous: the nature of the problem. *Earth Sci. Rev.* **19**, 305–338.

Barron, E.J. (1984) Climatic implications of the variable obliquity explanation of Cretaceous–Paleogene high-latitude floras. *Geology* **12**, 595–598.

Barron, E.J. & Peterson, W.H. (1990) Mid-Cretaceous ocean circulation: results from model sensitivity studies. *Paleoceanogr.* **5**, 319–337.

Barron, E.J. & Washington, W.M. (1982) Cretaceous climate: comparison of atmospheric simulations with the geologic record. *Palaeogeogr. Palaeoclim. Palaeoecol.* **40**, 103–133.

Barron, E.J. & Washington, W.M. (1985) Warm Cretaceous climates: high atmospheric CO_2 as a plausible mechanism. In: *The Carbon Cycle and Atmospheric CO_2: Natural Variations Archean to Present* (Eds E.T. Sundquist & W.S. Broecker) pp. 546–553. American Geophysical Union: Washington.

Barron, E.J., Arthur, M.A. & Kauffman, E.G. (1985) Cretaceous rhythmic bedding sequences: a plausible link between orbital variations and climate. *Earth Planet. Sci. Lett.* **72**, 327–340.

Berner, R.A., Lasaga, C. & Garrels, A.M. (1983) The carbonate–silicate geochemical cycle and its effect on atmospheric carbon dioxide over the last 100 million years. *Am. J. Sci.* **283**, 641–683.

Brass, G.W., Southam, J.R. & Peterson, W.H. (1982) Warm saline bottom waters in the ancient ocean. *Nature* **296**, 620–623.

Chaloner, W.G. & Creber, G.T. (1990) Do fossil plants give a climatic signal? *J. Geol. Soc. Lond.* **147**, 343–350.

Creber, G.T. & Chaloner, W.G. (1985) Tree growth in the Mesozoic and early Tertiary and the reconstruction of palaeoclimates. *Palaeogeogr. Palaeoclim. Palaeoecol.* **52**, 35–60.

Crowley, T.J. & North, G.R. (1991) *Paleoclimatology.* Oxford Monographs on Geology and Geophysics, 18, 339 pp. Oxford University Press: New York.

Crowley, T.J., Short, D.A., Mengel, J.G. & North, G.R. (1986) Role of seasonality in the evolution of climate during the last 100 million years. *Science* **231**, 579–584.

Crowley, T.J., Mengel, J.G. & Short, D.A. (1987) Gondwanaland's seasonal cycle. *Nature* **329**, 803–807.

Dettman, M.E. (1989) Antarctica: Cretaceous cradle of austral temperate rainforests? In: *Origins and Evolution of the Antarctic Biota* (Ed. J.A. Crame). Spec. Publ. Geol. Soc. Lond. 47, 89–105.

Ditchfield, P. & Marshall, J.D. (1989) Isotopic variation in rhythmically bedded chalks: paleotemperature variation in the Upper Cretaceous. *Geology* **17**, 842–845.

Douglas, J.G. & Williams, G.E. (1982) Southern polar forests: the Early Cretaceous floras of Victoria and their palaeoclimatic significance. *Palaeogeogr. Palaeoclim. Palaeoecol.* **39**, 171–185.

Douglas, R.G. & Woodruff, F. (1981) Deep sea benthic foraminifera. In: *The Sea* (Ed. C. Emiliani) pp. 1233–1327. Wiley Interscience: New York.

Epshteyn, O.G. (1978) Mesozoic–Cenozoic climates of northern Asia and glacial-marine deposits. *Int. Geol. Rev.* **20**, 49–58.

Fischer, A.G., De Boer, P.L. & Premoli Silva, I. (1990) Cyclostratigraphy. In: *Cretaceous Resources, Events and Rhythms* (Eds. R.N. Ginsburg & B. Beaudoin) pp. 139–172. Kluwer Academic Pulishers: The Netherlands.

Frakes, L.A. (1979) *Climates throughout Geologic Time.* Elsevier: Amsterdam.

Frakes, L.A. & Francis, J.E. (1988) A guide to Phanerozoic cold polar climates from high-latitude ice-rafting in the Cretaceous. *Nature* **333**, 547–549.

Frakes, L.A. & Francis, J.E. (1990) Cretaceous palaeoclimates. In: *Cretaceous Resources, Events and Rhythms* (Eds R.N. Ginsburg & B. Beaudoin) pp. 273–287. Kluwer Academic Publishers: The Netherlands.

Frakes, L.A., Francis, J.E., & Syktus, J.I. (in press) *Climates of the Phanerozoic.* Cambridge University Press: Cambridge.

Francis, J.E. (1986) Growth rings in Cretaceous and Tertiary wood from Antarctica and its palaeoclimatic implications. *Palaeontology* **48**, 285–307.

Francis, J.E. (1987) The palaeoclimatic significance of growth rings in late Jurassic/early Cretaceous fossil wood from southern England. In: *Applications of Tree-ring Studies* (Ed. R.G.W. Ward) pp. 21–36. British Archaeological Reports: Oxford.

Gradstein, F.M. & Sheridan, R.E. (1983) On the Jurassic Atlantic Ocean and a synthesis of results of DSDP Project Leg 76. *Initial Reports of the Deep Sea Drilling Project* **76**, 913–943.

Gregory, R.T., Douthitt, C.B., Duddy, I.R., Rich, P. & Rich, T.H. (1989) Oxygen isotope composition of carbonate concretions from the Lower Cretaceous of Victoria, Australia: implications for the evolution of meteoritic waters on the Australian continent in a paleopolar environment. *Earth Planet. Sci. Lett.* **92**, 27–42.

Hallam, A. (1984) Continental humid and arid zones during the Jurassic and Cretaceous. *Palaeogeogr. Palaeoclim. Palaeoecol.* **47**, 195–223.

Hallam, A. (1985) A review of Mesozoic climates. *J. Geol. Soc. Lond.* **142**, 433–445.

Hallam, A. (1987) Mesozoic marine organic-rich shales. In: *Marine Petroleum Source Rocks* (Eds J. Brooks & A.J. Fleet). Geol. Soc. Spec. Publ. 26, 251–261.

Hambrey, M.J. & Harland, W.B. (1981) *Earth's Pre-Pleistocene Glacial Record.* Cambridge University Press: Cambridge.

Haq, B.U. (1984) Paleoceanography: a synoptic overview of 200 million years of ocean history. In: *Marine Geology and Oceanography of the Arabian Sea and Coastal Pakistan* (Eds B.U. Haq & J.D. Milliman) pp. 201–231. Van Nostrand Reinhold Company: New York.

Hay, W.W. (1988) Paleoceanography: a review for the GSA Centennial. *Geol. Soc. Am. Bull.* **100**, 1934–1956.

Hudson, J.D. & Anderson, T.F. (1989) Ocean temperatures and isotopic compositions through time. *Trans. R. Soc. Edinburgh: Earth Sci.* **80**, 183–192.

Jefferson, T.H. (1982) Fossil forests from the Lower Cretaceous of Alexander Island, Antarctica. *Palaeontology* **25**, 681–708.

Jell, P.A. & Roberts, J. (1986) Plants and invertebrates from the Lower Cretaceous Koonwarra Fossil Bed, South Gippsland, Victoria. *Mem. Ass. Austral. Palaeontol.* **3**, 1–205.

Kemper, E. (1987) Das Klima der Kreide-Zeit. *Geol. Jahrb.* **A96**, 5–185.

Kolodny, Y. & Raab, M. (1988) Oxygen isotopes in phosphatic fish remains from Israel: paleothermometry of tropical Cretaceous and Tertiary shelf waters. *Palaeogeogr. Palaeoclim. Palaeoecol.* **64**, 59–67.

Krassilov, V.A. (1981) Changes in Mesozoic vegetation and the extinction of the dinosaurs. *Palaeogeogr. Palaeoclim. Palaeoecol.* **34**, 207–224.

Kutzbach, J.E., Guetter, P.J. & Washington, W.M. (1990) Simulated circulation of an idealized ocean for Pangaean time. *Paleoceanogr.* **5**, 299–317.

Parrish, M.J., Parrish, J.T., Hutchinson, J.H. & Spicer, R.A. (1987) Late Cretaceous vertebrate fossils from the North Slope of Alaska and implications for dinosaur ecology. *Palaios* **2**, 377–389.

Parrish, J.T. (1987) Global palaeogeography and palaeoclimate of the late Cretaceous and early Tertiary. In: *The Origins of Angiosperms and their Biological Consequences* (Eds E.M. Friis, W.G. Chaloner & P.R.Crane) pp. 51–73. Cambridge University Press: Cambridge.

Parrish, J.T. & Curtis, R.L. (1982) Atmospheric circulation, upwelling and organic-rich rocks in the Mesozoic and Cenozoic eras. *Palaeogeogr. Palaeoclim. Palaeoecol.* **40**, 31–66.

Parrish, J.T. & Spicer, R.A. (1988) Late Cretaceous terrestrial vegetation: a near-polar temperature curve. *Geology* **16**, 22–25.

Parrish, J.T., Ziegler, A.M. & Scotese, C.R. (1982) Rainfall patterns and the distribution of coals and evaporites in the Mesozoic and Cenozoic. *Palaeogeogr. Palaeoclim. Palaeoecol.* **40**, 67–101.

Pedersen, T.F. & Calvert, S.E. (1990) Anoxia vs. productivity: what controls the formation of organic-carbon-rich sediments and sedimentary rocks? *Am. Ass. Petrol. Geol. Bull.* **74**, 454–466.

Pirrie, D. & Marshall, J.D. (1990) High-paleolatitude Late Cretaceous paleotemperatures: new data from James Ross Island, Antarctica. *Geology* **18**, 31–34.

Rich, T.H. & Rich, P.V. (1989) Polar dinosaurs and biotas of the early Cretaceous of southeastern Australia. *Nat. Geogr. Res.* **5**, 15–52.

Saltzman, B. & Barron, E.J. (1982) Deep circulation in the Late Cretaceous: oxygen isotope paleotemperatures from *Inoceramus* remains in DSDP cores. *Palaeogeogr. Palaeoclim. Palaeoecol.* **40**, 167–181.

Schneider, S.H., Thompson, S.L. & Barron, E.J. (1985) Mid-Cretaceous continental surface temperatures: are high CO_2 concentrations needed to simulate above-freezing winter conditions? In: *The Carbon Cycle and Atmospheric CO_2: Natural Variations Archean to Present* (Eds E.T. Sundquist & W.S. Broecker) pp. 554–559. American Geophysical Union: Washington.

Sladen, C.P. (1983) Trends in early Cretaceous clay mineralogy in NW Europe. *Zitteliana* **10**, 349–357.

Sloan, L.C. & Barron, E.J. (1990) 'Equable' climates during Earth history? *Geology* **18**, 489–492.

Sloan, L.C. & Barron, E.J. (1991) Reply to 'Comments and reply on "equable' climates during Earth history?'. *Geology* **19**, 540–542.

Spicer, R.A. (1987) The significance of Cretaceous flora of Northern Alaska for the reconstruction of the climate of the Cretaceous. *Geol. Jahrb.* **96**, 265–291.

Spicer, R.A. (1990) Climate from plants. In: *Palaeobiology* (Eds D.E.G. Briggs & P.R. Crowther) pp. 401–403. Blackwell Scientific Publications: Oxford.

Spicer, R.A. & Parrish, J.T. (1986) Paleobotanical evidence for cool north polar climates in middle Cretaceous (Albian–Cenomanian) time. *Geology* **14**, 703–706.

Spicer, R.A. & Parrish, J.T. (1990) Late Cretaceous–early Tertiary palaeoclimates of the northern high latitudes: a quantitative view. *J. Geol. Soc. Lond.* **147**, 329–341.

Stevens, G.R. (1981) Relationship of isotopic temperatures and faunal realms to Jurassic–Cretaceous paleogeography, particularly of the south-west Pacific. *J. R. Soc. N. Zeal.* **1**, 145–158.

Vakhrameev, V.A. (1975) Main features of global phytogeography in the Jurassic and early Cretaceous. *Paleont. J.* **9**, 247–255.

Vakhrameev, V.A. (1981) Pollen *Classopollis*, indicator of Jurassic and Cretaceous climates. *The Palaeobot.* **28**, 301–307.

Veevers, J.J. & Ettriem, S.L. (1988) Reconstruction of Antarctica and Australia at breakup (95 ± 5 Ma) and before rifting (160 Ma). *Aust. J. Earth Sci.* **35**, 355–362.

Wing, S.L. (1991) Comments and reply on '"equable" climates Earth history?' *Geology* **19**, 539–540.

Wolfe, J.A. & Upchurch, G.R. (1987) North American non-marine climates and vegetation during the Late Cretaceous. *Palaeogeogr. Palaeoclim. Palaeoecol.* **61**, 33–77.

Ziegler, A.M., Hulver, M.L., Lottes, A.L. & Schmactenberg, W.F. (1984) Uniformitarianism and palaeoclimates: inferences from the distribution of carbonate rocks. In: *Fossils and Climate* (Ed. P.J. Brenchley) pp. 3–25. John Wiley & Sons: New York.

3 The recognition and stratigraphic implications of orbital-forcing of climate and sedimentary cycles

GRAHAM P. WEEDON

Introduction

The idea that cyclic changes in the nature of Earth's orbit have caused repeated climatic fluctuations (now known as the Milankovitch Theory) is not particularly new to geology. Herschel (1832) considered that changing eccentricity and precession would be important climatically over geological time-scales, but discounted the importance of obliquity variations (Table 3.1; Fig. 3.1). Adhémar (1842) was the first to argue that the Pleistocene ice ages were related to the orbital cycles, but Croll (1864) produced the first credible argument explaining how this could occur (Imbrie & Imbrie, 1979). In 1872 Lyell devoted 23 pages of *Principles of Geology* to an explanation and discussion of these ideas. Then Gilbert (1895) proposed that limestone–shale couplets in Colorado were related to the orbital-precession cycle and used them to estimate the duration of the late Cretaceous (Fischer, 1980). So began a series of papers considering an orbital–climatic explanation for sedimentary cyclicity. However, sedimentological and stratigraphic work on ancient cyclic sequences progressed largely independently of Pleistocene palaeoclimatological studies until the 1980s.

A primary concern of many areas of palaeoclimatological research is the reconstruction and modelling of particular climatic regimes. Thus, for instance, the occurrence of evaporites can be used to infer relatively arid conditions and this might be explained in terms of a particular palaeolatitude and/or topography (e.g. Hay *et al.*, 1982). Glacial striae on rock pavements and fossil tillites can be used to reconstruct the motion and extent of ancient ice sheets (Frakes, 1979). And palaeobotanical data can be used to determine mean annual temperatures (e.g. Spicer & Parrish, 1990). On the other hand, studies of orbital-climatic cyclicity start by concentrating on the nature of climatic *variability* rather than the climates themselves. In particular this variability is characterized by regularity and cycle periods of tens of thousands of years. Such characteristics can sometimes be inferred from stratigraphic logs, or time series, of cyclically varying sediment composition.

Studies of Pleistocene palaeoceanographic time series have revealed that a wide range of modern depositional systems respond to orbital-forcing in a wide range of climatic regimes. In addition, the methods now used to investigate time series from pre-Pliocene sections are often derived from work on the Pleistocene. Accordingly, this chapter starts by considering the Pleistocene precedents for examining ancient sequences for evidence for sedimentary cyclicity related to orbitally-forced climatic

Table 3.1 The nature and climatic effects of the orbital cycles

Orbital cycle	Current period (000 years)	Orbital effect	Climatic impact
Eccentricity	410, 123, 95	Degree of ellipticity of the orbit	Mainly controls size of precession effect; small control on total yearly radiative heating
Obliquity	41	Angle of axial tilt	Determines degree of seasonality
Precession	23, 19	Direction of axial tilt	Controls timing of seasons relative to perihelion and hence total radiative heating in each season

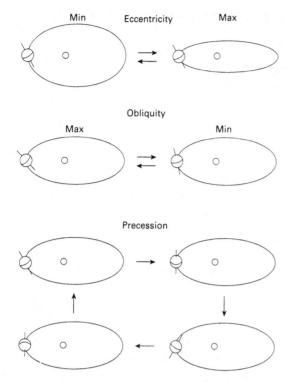

Min   Eccentricity   Max

Obliquity

Max   Min

Precession

Fig. 3.1 The nature of three orbital parameters which cause orbital–climatic cyclicity. All vary simultaneously, but here each is illustrated schematically at extremes of variation with the other elements kept constant. The sun is indicated by an open circle. In each illustration the Earth is shown at perihelion (closest point to the sun in each orbit or year). Climatic precession involves variation in both the direction of tilt of the Earth's axis (illustrated) and in the direction of the long axis of the orbital ellipse (not shown) which together determine the timing of perihelion relative to the seasons (Table 3.1).

changes. This is followed by a consideration of the depositional and environmental mechanisms that might link climate changes to sedimentary cycles. Then comes an examination of the methods used, and difficulties encountered, while trying to decide if orbital-forcing of climate was involved in the first place. In particular this involves discussion of the nature and pitfalls of time-series and power-spectral analysis. Finally, the problems and potential of interval dating based on counts of regular cycles is addressed.

The high-resolution of the records of varying sediment composition generated in this field of research are of great value stratigraphically — as emphasized throughout this chapter. This chapter

also considers, at appropriate points, some of the implications of the new ideas emerging from study of non-linear dynamical systems — known as 'Chaos Theory'. A useful introduction to this subject is provided by Stewart (1990) and more advanced treatment is given by Schuster (1984). It appears that many aspects of climatic change might be explicable in terms of chaotic behaviour. Here, where non-linear or chaotic processes are mentioned, it is worth remembering that the field is still very young and requires much more investigation. However, these ideas have been included because they help understand the nature of the background climatic changes from which the orbital–climatic variability needs to be distinguished.

Pleistocene studies and methods

The orbital variations themselves (Table 3.1; Fig. 3.1) result from the changing gravitational environment of the Earth as dictated by the positions of the other planets (Berger, 1984). Because the system evolves in a vacuum it is almost undamped so that the cycle periods remain almost constant (strictly each orbital cycle is 'quasi-periodic'). The orbital cycles have the effect of changing the average distribution of incident radiation at the top of the atmosphere with cycle periods of tens to hundreds of thousands of years. Croll's (1864) speculations on a link between orbital variations and climatic changes could not be tested satisfactorily immediately so his ideas were left in limbo for some time (as reviewed by Imbrie & Imbrie, 1979).

Milankovitch (1941) pioneered detailed calculation of the changes in insolation distribution in the past (Imbrie & Imbrie, 1979; Milankovitch, 1984; Berger, 1988). Current work in this area is partly concerned with extending the results back in time and with refining the numerical solutions by reducing the number of simplifications used for this dynamic nine-body problem (Berger, 1984). At present it appears that the calculated insolation curves (Fig. 3.2) are reliable back to 5 Ma BP.

However, for the treatment of times earlier than this Laskar (1989) demonstrated that the solar system may have behaved chaotically in terms of orbital trajectories. Nevertheless, it appears that the frequencies of the orbital cycles evolved predictably (Berger et al., 1991). This means that although the precise relationship between different orbital cycles at a particular instant cannot be determined for a point in the Mesozoic, for example, the frequencies

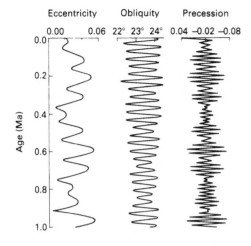

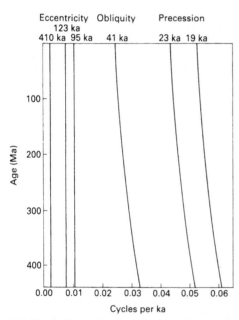

Fig. 3.2 The orbital cycles through time. *Top* — a plot of the variations of the orbital elements over the last 1 Ma shows that the different orbital cycles have characteristic periods. The impact on insolation (radiation flux) of these variations for the different orbital parameters is not in proportion to the amplitudes shown. In fact the eccentricity cycle has a very small direct impact on changes in insolation. However, eccentricity does exert a major control on the amplitude of the precession cycles and hence on precession-related changes in insolation. *Bottom* — the frequencies of the precession and obliquity cycles gradually became lower through geological time due to changes in the Earth–Moon distance (from Berger *et al.*, 1989). The *current* main orbital cycle periods are indicated at the top of this plot.

of the orbital cycles can be established (Berger *et al.*, 1989). In fact changes in the Earth–Moon distance mean that the lengths of the precession and obliquity cycles became slightly longer through time (Fig. 3.2).

Milankovitch also suggested that the link between orbital changes and ice-sheet volumes was determined by the occurrence of cool summers rather than severe winters. Palaeoclimatic models have developed considerably since: it is now thought that this link is indirect and that ice-sheet volumes are also controlled by mechanisms that include slow ice-sheet growth and rapid decay (Imbrie & Imbrie, 1980; Ruddiman & McIntyre, 1981; Berger, 1988). It also appears that variations in atmospheric concentrations of CO_2 are forced by changes in insolation distribution so amplifying global temperature changes (Genthon *et al.*, 1987; Shackleton & Pisias, 1985). Such a process could help explain the synchroneity of ice-volume changes in both the north and south hemispheres.

Emiliani (1955, 1966) used oxygen isotope measurements from Foraminifera from deep-sea cores as a measure of climatic state. He believed that the observed variations in isotopes mainly reflected sea-surface temperature changes, but it now appears that global ice volumes affected seawater composition significantly and that this provided a much greater control on oxygen isotopic values in the Pleistocene (Shackleton & Opdyke, 1973). The similarity of isotope variations from several cores was taken as a climatic indicator and Emiliani (1966) suggested that there was a very close correlation with calculated insolation changes. However, it was not until the mid-1970s that sufficiently long isotope records and reliable radiometric dating of magnetic reversals permitted a rigorous comparison of the calculated insolation variations and the isotopic records. Nevertheless, Emiliani pioneered the production of palaeoclimatic time series from deep-sea cores which is now a routine part of Pleistocene palaeoceanography (Ruddiman, 1985).

Hays *et al.* (1976), in a key paper, dated down-core measurements of $\delta^{18}O$, estimates of sea-surface temperatures and faunal composition for the last half million years by assuming constant sedimentation rates between a few dated horizons. Power-spectral analysis detected regular cyclicity with periods of 100 000, 41 000, 23 000 and 19 000 years which matched the orbital cycle periods. Filtering the time series showed that amplitude variations of each regular cycle matched, but lagged, the calculated insolation cycles. Next a constant lag of the measured isotope ratios (ice-volume index) behind the calculated insolation cycles was used to produce a record of $\delta^{18}O$ 'tuned' to the orbital cycles. This was designed to remove the effects of short-term (1000–100 000 year) variations in sedimentation rate and allow production of a standard oxygen isotope record. The tuning method has now been applied to a large number of cores and the standard record of oxygen isotope stages has now been extended back to 2.5 Ma BP (Imbrie *et al.*, 1984; Ruddiman *et al.*, 1989; Raymo *et al.*, 1989; Shackleton *et al.*, 1990).

In Fig. 3.3 bottom-water oxygen isotope variations are illustrated for the last 2.5 Ma for sites in the North Atlantic (DSDP 607) and East Pacific (ODP 677) using the time-scale adopted by Shackleton *et al.* (1990). It is apparent that although the two records have different sampling densities and are obtained from different oceans, they are extremely well-correlated. The size and width of the peaks, labelled '41 ka' on the accompanying power spectra, demonstrate that global ice-volume variations related to the obliquity cycle are large and very regular in these records. The '100 ka' peaks, however, are smaller reflecting a smaller contribution

to the records. In fact the 100 000-year variations in ice volumes were minor prior to 0.9 Ma BP (Ruddiman *et al.*, 1989).

The origin of the late Pleistocene 100 000-year cycles in ice volume remains controversial and difficult to model (e.g. Hays *et al.*, 1976; Imbrie & Imbrie, 1980; Le Treut & Ghil, 1983; Berger, 1988). These cycles are very considerably larger in amplitude in the late Pleistocene than would be expected from a linear model of orbital-forcing (Hays *et al.*, 1976). One of the many possibilities is that during a change in the nature of the climate system (Ruddiman *et al.*, 1989), internal, unforced natural oscillations in the climate with periods near 100 000 years became phase-locked and 'resonated' with the regular 100 000-year orbital-forcing — typical behaviour for a non-linear system (Saltzman & Maasch, 1990; Stewart, 1990).

When the $\delta^{18}O$ values of Fig. 3.3 are plotted against core-depth rather than time (Fig. 3.4), there still appears to be a cyclicity. Yet the corresponding power spectra show that variations in isotopic ratios with depth are actually much less regular. This reflects a distortion of the original signal caused by irregular variations in sedimentation rates over periods of thousands to a few hundred thousand

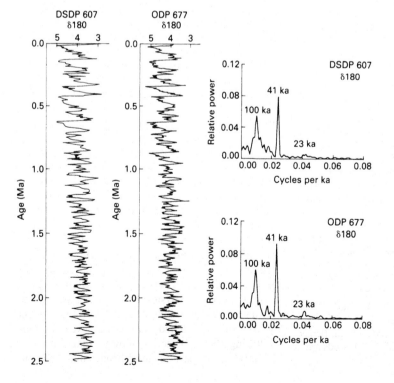

Fig. 3.3 Oxygen isotope records from benthonic Foraminifera for the North Atlantic (DSDP Site 607) and the East Pacific (ODP Site 677) covering the last 2.5 Ma (Raymo *et al.*, 1989, Ruddiman *et al.*, 1989, Shackleton *et al.*, 1990). The similarity of the records reflects a control of bottom-water isotopes by global ice volumes (Shackleton & Opdyke, 1973). The power spectra show that these records are dominated by regular cycles with periods matching those of the orbital cycles. All the spectra of this paper have been generated using the standard Blackman–Tukey method. Relative power (strictly relative variance density) values were obtained by dividing all the power values by the sum of these values.

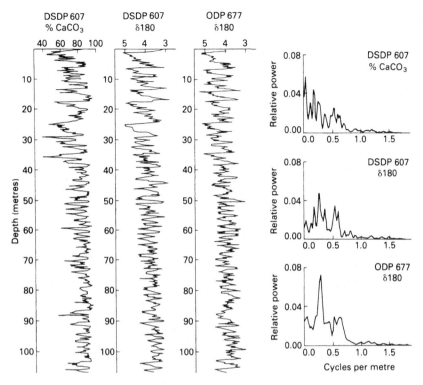

Fig. 3.4 The same isotope data as in Fig. 3.3, but with values plotted as a function of depth rather than time. Carbonate contents for DSDP Site 607 are also illustrated. By chance the sedimentation rates at the two sites were almost identical. The similarity of the isotope (proxy-ice volume) and carbonate records shows that Pleistocene sedimentary cycles can act as proxy-records of palaeoclimatic variation.

years. Also plotted in Fig. 3.4 are calcium carbonate contents for the North Atlantic site. These values do not have oscillations of the same shape as the oxygen isotopes, but the same cyclicity is present. This type of broad agreement between palaeoclimatic indicators and bulk sediment composition provides the Pleistocene precedent for examining ancient deep-sea sequences for signs of orbital-forcing.

There are many types of sedimentological model used to explain the link between Pleistocene orbital variations and sedimentary cycles. Most have been developed for deep-sea sediments because shelf sediments rarely accumulated continuously due to the large sea-level fluctuations (Fairbanks, 1989). Some of these models involve ice-sheet processes as for example in the North Atlantic. Their variations in bulk composition (%CaCO$_3$, Fig. 3.4) are mainly related to dilution of the biogenic carbonate by ice-rafted siliciclastic material, although variations in carbonate productivity and dissolution probably

had some impact (Ruddiman *et al.*, 1989). On the other hand, in the East Pacific it appears that variations in the depth and intensity of sea-floor dissolution, perhaps partly linked to atmospheric CO$_2$ concentrations, played the major role in producing cyclic variations in carbonate contents (Farrell & Prell, 1989).

In the NW Indian Ocean, cycles in opal content are related to variations in productivity as determined by the intensity of upwelling which is controlled, in turn, by the strength of the monsoon wind (Murray & Prell, 1991). However, cycles in carbonate content are related to variable dilution by wind-blown terrigenous clastics and diagenetic dissolution. In the same sediments organic-carbon contents vary as a function of preservation linked to sedimentation rates (Murray & Prell, 1991). For the Pleistocene sapropels in the East Mediterranean Rossignol-Strick *et al.* (1982) and Rossignol-Strick (1983) argued that variation in organic-carbon preservation was linked to changes in productivity and density stratification. In this model it was suggested that insolation in Africa determined the intensity of the monsoon rainfall and thus runoff from the Nile into the Mediterranean.

Ancient sedimentary cycles and orbital–climatic controls

Examples of ancient sedimentary cyclicity attributed to orbital-forcing have now been described from the Precambrian and every period in the Phanerozoic. The wide range of facies involved, and some of the sedimentological models linking orbital variations to sedimentary cycles, are listed in Table 3.2. Comprehensive reviews of the various facies and the models are provided by Einsele (1982), Arthur et al. (1984), Fischer et al. (1985), Fischer (1986), Berger (1989) and Fischer et al. (1990). Examples of cyclic sequences are illustrated in Fig. 3.5.

Producing descriptive models to explain how climatic change caused sedimentary cyclicity with or without the intervention of ice is not problematic and is often based on the Pleistocene examples given earlier. However, as discussed below, choosing the correct climatic and sedimentological processes in each case is difficult. Quantitative modelling of cyclic sequences has also been attempted (e.g. Read et al., 1986; Walkden & Walkden, 1990). This has mainly been used as a form of experimentation with different input parameters in order to examine the nature of the output. This approach has certainly provided some useful insights into the build-up of carbonate platforms and clastic cyclothems.

The asymmetry of many shallow-water cyclic sequences has been modelled using asymmetric changes in relative sea-level. This has arisen from the observation that late Pleistocene sea-level changes, as deduced from oxygen isotope curves, appear to have been highly asymmetric. However, it is clear from Fig. 3.3 that this asymmetry starts in the late Pleistocene with the arrival of major 100 000-year cycles. As these cycles may owe their existence to the particular boundary conditions of the time (Ruddiman et al., 1989), it is far from certain that asymmetric 100 000-year sea-level changes are an appropriate input for models of

Table 3.2 Examples of models linking regional or global climatic variations to sedimentary cycles

Facies	Link between climatic and sedimentary cycles	Example
Evaporitic	Aridity/humidity variation controls evaporite accumulation	Anderson, 1982, 1984
Lacustrine	Runoff and humidity controls lake depth and hence micro facies	Olsen, 1986
Clastic shoreline	Position of shoreline related to glacioeustatic sea-level	Clifton, 1981
Deltaic	Position of delta front related to glacioeustatic sea-level	Van Tassell, 1987
Platform carbonate	Exposure, submergence and growth related to glacioeustatic sea-level	Schwarzacher & Haas, 1986
Shelf carbonate	Microfacies related to water depth and glacioeustatic sea-level	Schwarzacher, 1989
Siliciclastic shelf	Grain size dependent on turbulence under control of storminess or sea-level	Van Echelpoel & Weedon, 1990
Hemipelagic	%$CaCO_3$, e.g. related to clastics from runoff % organic carbon related to productivity and/or water column stratification	Weedon, 1985
Pelagic	%$CaCO_3$ related to productivity and/or supply of clastics and/or sea-floor dissolution	Herbert & Fischer, 1986
Turbiditic	Clastic supply related to sea-level in relation to shelf or carbonate platform	Foucault et al., 1987

Fig. 3.5 Examples of cyclic sequences related to orbital–climatic cyclicity. (a) Alternating nanofossil ooze and marly nanofossil ooze from Unit II, Upper Miocene of Owen Ridge, NW Indian Ocean (Prell *et al.*, 1989). This Ocean Drilling Program core is cut into 1.5 m sections. (b) Eight metres of silts and clays from the Boom Clay Formation, Lower Oligocene, Belgium (Van Echelpoel & Weedon, 1990). (c) Nine metres of light- and dark-coloured marls and laminated shales from the Belemnite Marls, Lower Jurassic, England (see Fig. 3.6; Weedon & Jenkyns, 1990). (d) Three metres of limestones, marls and shales of the Scisti a Fucoidi, mid-Cretaceous, Italy (Herbert & Fischer, 1986).

pre-Pleistocene cyclothems and carbonate platform sequences. Indeed the simulations of Walkden & Walkden (1990) demonstrate that asymmetric sea-level variations are not essential for generating asymmetric shallow-water sedimentary cyclicity.

Good examples of descriptive models are provided by the work on the cyclic Bridge Creek Member of the Greenhorn Formation from the Albian to mid-Turonian (Cretaceous) of the West-

ern Interior Seaway of North America. Gilbert (1895) originally suggested that dilution cycles were involved with clay transported from land and the carbonate precipitated in the sea. Barron *et al.* (1985), on the basis of petrographic, trace-fossil and geochemical observations, argued that orbitally-driven changes in precipitation and runoff led to changes in clay fluxes and produced carbonate dilution cycles. Periodic density stratification, caused by the formation of a brackish-water 'lid' over the Seaway was thought to account for periodic bottom-water anoxia and increased organic-carbon preservation. In particular, Barron *et al.* (1985) proposed that variations in the intensity of summer monsoon rainfall and winter storminess would have been critical. However, when the impact of changes in radiation budgets due to orbital changes was modelled, using a General Circulation Model for the Cretaceous, this failed to generate any monsoonal precipitation —though winter storminess did vary (Glancy *et al.*, 1986). It is not clear whether this reflects the inadequacy of the model or a reliable result.

Eicher & Diner (1989) argued that faunal diversities and lateral variations in organic carbon preservation do not support the brackish lid model. Instead they favoured productivity variations related to changes in oceanic and seaway circulation patterns with higher productivity during limestone deposition. Numerical hydrodynamic modelling suggests that seaway circulation was dominated by storms (Erickson & Slingerland, 1990). Additionally, on the basis of nanofossil assemblages, although productivity variations can be inferred, fertility was apparently highest during marl, rather than limestone, deposition (Watkins, 1989).

The Bridge Creek Member has been investigated on-and-off for nearly a century using a great many techniques, but choosing the correct climatic and sedimentological model remains controversial. This observation suggests that searching for the 'correct' environmental model for any sequence is highly problematic and time-consuming (though of course far from pointless). A more immediately rewarding approach is to use the detailed stratigraphic information generated in such studies to provide very high-resolution logs. These data may be invaluable for correlation and interval dating — regardless of the environmental models advanced in explanation (see below). First, it is essential to demonstrate that orbital–climatic control of cyclicity occurred.

To date, in the majority of cases where sedimentary cyclicity has been related to orbital-forcing, spectral analysis has not been employed. Instead cycles between approximately dated levels have been counted to arrive at average periods. This often suffers from difficulties in defining what constitutes a 'cycle', because there can be no distinction between regular and irregular variations in composition. Additionally, it is difficult distinguishing between two or more regular cycles which are

superimposed on each other. Because regularity is not established in this approach, other irregular environmental factors must be considered as well (e.g. Cisne, 1986; Goldhammer *et al.*, 1990). Further, Algeo & Wilkinson (1988) noted that in many shallow-water environments net accumulation rates have a small range because they are ultimately limited by rates of crustal subsidence. This means that for tidal, deltaic and subtidal environments, cycles that have wavelengths of 1–20 m usually have average periods calculated at between 20 000 and 400 000 years regardless of the mode of origin (orbital–climatic or otherwise).

Spectral analysis provides a more objective method for testing orbital–climatic control because regular cyclicity can be looked for, and superimposed regular cycles can be distinguished from each other and from noise. In fact spectral analysis was applied for this purpose well before Pleistocene palaeoclimatologists widely accepted the viability of the Milankovitch Theory (Schwarzacher, 1964). Indeed regular cyclicity has been demonstrated with power spectra in every facies where orbital–climatic control has been suggested (see examples in Table 3.2).

Of course in pre-Pliocene sequences dating is rarely sufficiently accurate to permit use of the dating method of Hays *et al.* (1976) (i.e. specifying the ages of certain horizons and creating a time-depth model). It is also not yet possible to calculate the exact orbitally-driven changes in insolation prior to about 5 Ma BP (Berger, 1984). However, although the periods of some of the orbital cycles were smaller in the past (Fig. 3.2), in any interval spanning just a few million years they were as regular as in the last few million years (Berger *et al.*, 1989). Thus using time series of rock composition against stratigraphic height, spectral analysis can be used to find regular cycles in thickness. Crude dating is then employed to decide if particular regular cycles have periods near to the orbital periods. The problems with this approach are discussed in the next two sections.

Generating time series

Power-spectral analysis is performed on successive values of some parameter obtained at constant intervals of time or space. The collection of values is called a 'time series' even if the values were obtained as a function of distance rather than time (Priestley, 1981). Likewise the horizontal axis of the resulting power spectrum is the 'frequency axis' even if calibrated in units of cycles per metre. Before spectra are generated the average value of the series is subtracted from all the measured values so that the oscillations occur around the mean or zero line. The definition of geological cyclicity has been reviewed by Schwarzacher (1975). Here 'sedimentary cycles' and 'cyclic sequences' refer to *any* sort of time series oscillations associated with the interbedding of two or more different rock types. 'Regular cycles' is reserved for oscillations that have a near constant wavelength. Before time series are generated from ancient cyclic sequences, several procedural decisions are needed. These concern the parameter to record, the sample spacing to use and the number of values required.

Proxy-palaeoclimatic parameters

In studies of the Pleistocene it is possible to generate palaeoclimatic records by, for instance, calculating past sea-surface temperatures based on assemblages of planktonic Foraminifera whose

Table 3.3 Parameters used for time-series generation in ancient cyclic sequences

Parameter	Example
%CaCO$_3$	Herbert & Fischer, 1986
% Organic carbon	Weedon & Jenkyns, 1990
Si/Al	Herbert et al., 1986
% Total salt	Williams, 1991
$\delta^{18}O, \delta^{13}C$	Ditchfield & Marshall, 1989
Fossil assemblage index	Cottle, 1989
Fossil abundance index	Cottle, 1989
Grain size	Van Echelpoel & Weedon, 1990
Magnetic inclination/ declination	Napoleone & Ripepe, 1989
Magnetic susceptibility	Weedon et al. (in preparation)
Bed thickness index	De Boer, 1982
Fischer plot	Read & Goldhammer, 1988
Rock type index	Schwarzacher, 1987a
Colour densitometry	Herbert & Fischer, 1986
Gamma ray log	Laferriere & Hattin, 1989

modern temperature tolerances are known. In ancient sequences only proxy-palaeoclimatic records are available. Clearly it is essential, especially in pelagic and hemipelagic sequences, to consider the impact of diagenesis on the supposed palaeoclimatic proxy (Weedon, 1991). Fortunately, it is possible in many cases to use burrow mottling and primary textural features such as lamination to decide whether diagenesis has enhanced or entirely masked primary compositional variations.

Event beds must also be considered. As explained later, spectral analysis is not dependent upon an assumption of identical sedimentation rate for different rock types. Nevertheless, it would be nonsense to treat event beds in the same fashion as the intervening more uniformly and more slowly accumulated material. However, the interval between event beds may be a crude indicator of the time between events, and the composition and thickness of event beds might, in some cases, provide useful palaeoclimatic information (Foucault et al., 1987; Haak & Schlager, 1989).

An indication of the range of parameters that have been used for time-series generation is provided in Table 3.3. Simultaneous measurements of a variety of different parameters from the same unit provides the opportunity to investigate the interrelationship of different features of the sedimentary environment using the cross-spectral methods described later.

Sample intervals and numbers of points

The choice of sample interval is of critical importance. If samples are collected too far apart then small-scale variations are incompletely recorded and spurious long-wavelength oscillations are created. Such incorrectly collected data are described as 'aliased' and any spectral analysis of such data cannot be trusted (Priestley, 1981; Pisias & Mix, 1988). Fortunately, bioturbation, by mixing different layers, smooths out very small-scale compositional variations. Thus, except in non-bioturbated sediments, it is rarely necessary to collect samples less than a centimetre apart. The sample interval needed can be determined with a pilot study using very closely-spaced samples (e.g. Van Echelpoel & Weedon, 1990).

The frequency resolution of power spectra depends on the sample interval and the total number of points in the series. For a certain sample interval, more points gives a higher frequency resolution. As most procedures of spectral analysis require values at uniform spacing, irregularly spaced points require interpolation. The interpolation interval should be chosen to be larger or equal to the average sample interval so that the interpolated data are not biased. Equally spaced data are always preferable. Generally, if regular cycles of a particular wavelength are suspected, the time series needs to be at least 5–10 times that length for their detection spectrally (Schwarzacher, 1975). Naturally the longest obtainable time series is desirable, but most series are limited by the length of continuous exposure.

Spectral analysis and its pitfalls

A simple introduction to the spectral analysis of stratigraphic time series and the mathematical jargon involved has been provided elsewhere (Weedon, 1991). Priestley (1981) provides a detailed mathematical treatment.

Many types of spectral mathematical procedures are available for any time series and all have their own advantages and limitations (Pestiaux & Berger, 1984). The most widely used methods are based on

Fourier analysis of time series in terms of component sine and cosine waves. The power spectrum plots the squared average amplitude (i.e. the power) of each regular component sine and cosine wave against frequency. The size of a spectral peak indicates the *average* importance of that component to the whole series. If a particular regular cycle varies in amplitude along a time series, this has no impact on the spectrum as average amplitudes only are involved. For example, the orbital-precession cycle varies in amplitude through time due to the 410 000- and 100 000-year variations in eccentricity (Fig. 3.2). These modulations of precession cycle amplitude *do not* generate 410 000- and 100 000-year spectral peaks.

The shapes of power spectra and the significance of spectral peaks

Fourier analysis of a real-time series is based on a finite number of points. As a result, the calculated spectrum is always an approximation to the spectrum that would be derived for an infinitely long series. In fact the estimated spectrum contains unwanted peaks and troughs that result from both

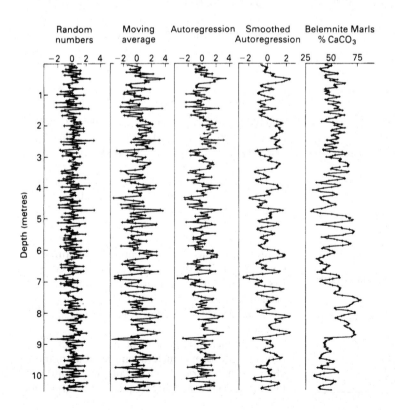

Fig. 3.6 Artificial noise time series and %CaCO$_3$ data from the Belemnite Marls. See text for the origin of the noise time series. All the series have 350 points and Fig. 3.7 shows the corresponding power spectra.

having a limited number of data points and from the effects of the data starting and stopping abruptly. To some extent both these problems can be alleviated during the calculation of the spectrum (Weedon, 1991). Yet unwanted peaks and troughs are always present as illustrated in Figs 3.6 and 3.7. In these figures artificial noise-time series and their spectra are compared with results for %CaCO$_3$ data obtained from the Belemnite Marls (Lower Jurassic, southern Britain). In all these cases Fourier spectra have been calculated for 350 points spaced at 3 cm intervals.

The first three times series in Fig. 3.6 are obtained from the appendices of Priestley (1981). The first is a set of random numbers with unit variance and the mean subtracted. All possible wavelengths of variation are equally represented in random number series so that the theoretical spectrum is a horizontal line. In Fig. 3.7 the peaks and troughs result from the finite length of the series, but the background levels are flat as expected. By analogy with white light, such a near-uniform contribution of all frequencies to the data is described as 'white noise'.

The second series was generated by Priestley (1981) using a three-point moving average applied to the first series. This averaging makes each successive value partly related to the previous value so there is some 'memory' in the data (also known as a Markov property, Schwarzacher, 1975). Because

very short-term oscillations have been smoothed-out by the averaging, higher frequency oscillations have smaller average amplitudes so the corresponding spectrum in Fig. 3.7 slopes to the right. This is described as 'red noise' because of the dominance of low-frequency oscillations (more correctly 'coloured noise', strictly 'red noise' is used to refer to one particular distribution of power with frequency).

In the third series each new value was calculated by adding a random number from the first series, to a proportion (0.6) of the previous random number (i.e. this is a first-order autoregressive series). Again this effectively acts as a form of smoothing of the first series and a red-noise spectrum results. Red-noise spectra and time series possessing a Markov property are apparently common to all sorts of stratigraphic time series (Schwarzacher, 1975). This is partly because real geological systems have 'inertia'. An instantaneous change in basin catchment area for instance, does *not* cause an instantaneous change to a new equilibrium state in terms of sedimentary processes within the basin.

Over intervals of thousands to millions of years, the spectrum of palaeoclimatic variations themselves is essentially red noise with peaks related to the orbital cycles superimposed on it (Shackleton & Imbrie, 1990). This means that aside from the regularities, the longer the interval considered, the larger the climatic changes are likely to be. Modelling of the impact of orbital-forcing on a non-linear

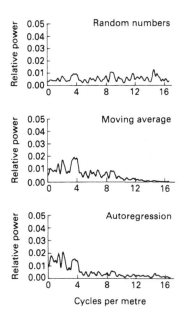

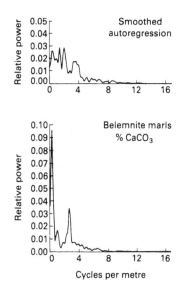

Fig. 3.7 Spectra for the time series of Fig. 3.6. Smoothing the random number data using a moving average or autoregressive process produces time series with the characteristics of 'red noise'. The smoothed autoregressive time series looks similar to the Belemnite Marls carbonate data in Fig. 3.6. However, the corresponding spectra show that the former still possesses a simple red noise spectrum whereas the geochemical data is dominated by two well-defined frequencies of variation superimposed on a red noise continuum.

climate system has successfully reproduced the general form of the climatic spectrum (Le Treut & Ghil, 1983). Unlike linear systems, in some non-linear systems it is possible for small amplitude, short-term disturbances to grow into large amplitude, long-period oscillations (James & James, 1989). This helps account for the red noise nature of the climate system.

The fourth time series of Fig. 3.6 was created by two applications of a weighted moving-average applied to the third series. This was designed to produce a series that looks similar to the %CaCO₃ data. Note that although the fourth series appears to have regular cyclicity, the corresponding spectrum shows that this is not the case (Fig. 3.7). (Note that in red-noise spectra the size of the background peaks and troughs increases towards low frequencies.) Conversely, the spectrum of the %CaCO₃ series is clearly dominated by two spectral peaks that relate to decimetre-scale couplets of light and dark marl and to metre-scale bundles of couplets (Fig. 3.8; Weedon & Jenkyns, 1990). These examples demonstrate how useful spectral analysis is for detecting regular cyclicity, and demonstrate that the visual inspection of time series is less objective than one might think.

In the case of the Belemnite Marls data, the spectral peaks are clearly visually distinct from the red-noise background. A more objective method for distinguishing real peaks from background variation in the spectrum is to use confidence levels applied to a log power vs. frequency plot (Schwarzacher, 1975,

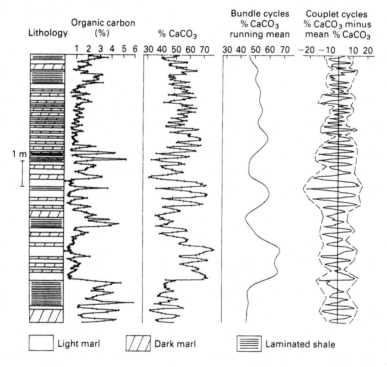

Fig. 3.8 Organic carbon and calcium carbonate data for the Belemnite Marls. The top 9 m of this section is illustrated in Fig. 3.5. The bundle cycles have been isolated by repeated smoothing of the carbonate data using a weighted three-point moving average (90 applications). This was used to remove all high-frequency components of the time series. The couplet cycle was then isolated by subtracting the smoothed values from the original data. The amplitude of the precession cycle is dependent on the eccentricity cycle (Fig. 3.2). If the bundles of the Belemnite Marls recorded eccentricity cycles, and the couplets recorded precession, a similar relationship would exist — even if some couplets were missing at undetected gaps. However, the filtering reveals that there is no relation between the amplitude of the couplet cycles and the shape of the bundle cycle. The bundle cycle determines the occurrence of laminated shales so it appears to be a primary, not a diagenetic phenomenon. It may record long-term irregular climatic variations. The more regular couplets probably record the orbital-precession cycle. A decrease in sedimentation rate towards the top of the section may explain the thinner couplets there.

pp. 209–212; Priestley, 1981). In the spectrum for the Belemnite Marls (Fig. 3.7), the couplet peak is significant at the 98% level and the bundle peak at the 90% level. For many spectral methods confidence levels cannot be calculated (Pestiaux & Berger, 1984). Instead the time series is split into two halves and a check that the same peaks are present in both subseries is made.

The effect of geological distortions

As a thickness rather than a time axis has to be used for pre-Pliocene cyclic sequences, several geological processes can 'distort' the environmental signal being investigated (Weedon, 1991). These processes destroy rather than create regularity. Therefore if a regular cyclicity is detected using a thickness scale, it can be concluded that a regular cycle in time was the cause (Schwarzacher, 1975).

Bioturbation tends to smooth compositional time series thereby increasing the slope of power spectra (compare results for the unsmoothed and smoothed autoregressive time series of Figs 3.6 and 3.7). Systematic variations in sedimentation rate and compaction which are related to sediment composition both have the same effect on time series. A sine-wave environmental signal becomes distorted into a cuspate or even a square-wave form which generates harmonic peaks on power spectra (Weedon, 1989). Harmonic peaks are readily identified as they lie at integer multiples of the frequency of the fundamental peak. Sometimes a strongly non-linear relationship between time and depth can generate combination tone peaks which are also readily recognized. Note that harmonic and tone peaks only occur in association with spectral peaks recording a basic regular cyclicity.

Short-term irregular variations in sedimentation rates and compaction distort an original sine-wave-like signal such that successive oscillations vary in wavelength. This leads to a broadening of spectral peaks (e.g. compare Figs 3.3 and 3.4). Trends in sedimentation rates, such as a monotonic increase up-section, can distort a cyclic signal so profoundly that no well-defined spectral peak is generated. In this context a 'sliding window' is used so that spectra for overlapping subsections of the data are generated (i.e. 'evolutionary spectra' of Priestley, 1981). These subspectra can then be compared to see whether any well-defined spectral peaks shift frequency up-section due to changing mean sedimentation rates (Melnyk & Smith, 1989).

Finally, the impact of hiatuses on the power spectra of stratigraphic time series depends upon the average spacing and size of the gaps. Stratigraphic gaps generate additional noise on spectra and can alter the wavelength ratios of pairs of regular cycles (Weedon, 1989). This has important consequences for estimating the periods of regular cycles (see below). Thus significant spectral peaks can be taken as a demonstration of regular cycles in time. But the *absence* of significant spectral peaks could indicate that either there was no regular cycle in the environmental signal or that time-depth distortions have removed the original evidence for regularity.

Filtering and cross spectra

Once regular cycles have been identified it is often useful to be able to isolate them by filtering. There are many methods of filtering, but essentially all involve reducing the number of frequency components considered together (Priestley, 1981). For instance 'low-pass' filtering involves removing all high-frequency components from a time series and 'band-pass' filtering allows the examination of oscillations in the data to be restricted to a small range of wavelengths. An example of filtering is shown in Fig. 3.8, where a weighted three-point moving average was applied repeatedly to the Belemnite Marls carbonate data.

As mentioned earlier at least 5–10 repetitions of a particular wavelength are needed to be able to claim confidently that a regular cycle has been identified. For example, the spectral peak associated with bundles of couplets corresponds to a wavelength of 300 cm even though only 3.5 repetitions of this wavelength could be present in a series 10.5 m long. In fact the wavelength resolution of the spectrum at this frequency, as defined by the frequency bandwidth of Fig. 3.9, means that this peak could relate to oscillations between 190 and 700 cm long. Indeed the filtering (Fig. 3.8) demonstrates that the bundles are actually rather irregular in wavelength. Additionally, the couplets clearly become abruptly thinner in the top-third of the data, presumably due to a decrease in mean sedimentation rate (Weedon & Jenkyns, 1990). However, the couplets in the lower two-thirds of the series are sufficiently constant in thickness to generate a spectral peak associated with a mean wavelength of 37.5 cm. At this part of the spectrum the bandwidth indicates that this peak relates to oscillations between 35 and

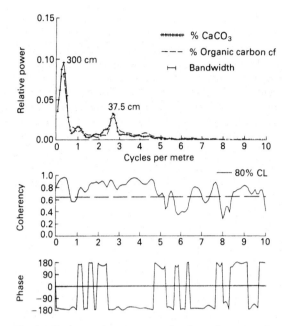

Fig. 3.9 Power- and cross-spectra for the carbonate and organic carbon time series of the Belemnite Marls. The organic carbon values were re-expressed on a carbonate-free basis to remove spurious correlations with the carbonate contents. The coherency spectrum reveals significant linear correlation between the two time series at the scales of the couplet and bundle cycles. The phase spectrum shows that at the same frequencies there is a phase difference of about 180°, i.e. an inverse relationship.

40 cm. Thus the couplets in the lower two-thirds of the data can justifiably be described as regular.

Filtered time series allow visual examination of amplitude variations up-section. Crude dating in the case of the Belemnite Marls suggests that the bundles had an average period near 100 000 years and the couplets near 20 000 years ago. The regularity and average period of the couplets suggests that the precession cycle is involved. However, if the bundles recorded eccentricity variations, then the amplitude variations of the couplets should be related to the shape of the bundle cycle (Fig. 3.2). This relationship would hold whether or not the section contains stratigraphic gaps. In fact in Fig. 3.8 it is clear that there is no such relationship. Instead the bundles have been ascribed to large, *irregular* variations in climate with periods of a few hundred thousand years (Weedon & Jenkyns, 1990). Such variations in climate have been described in Pleistocene data (Ruddiman *et al.*, 1989)

and may relate to a strongly non-linear or 'chaotic' climate system and represent part of a red-noise climatic spectrum.

Although filtering allows regular cycles to be compared visually in two time series, an alternative method is to generate cross spectra (Priestley, 1981). Coherency spectra are used to test whether amplitude variations in two data sets are linearly correlated at different frequencies. Similarly phase spectra indicate whether, at particular frequencies, two time series vary synchronously or are offset (Fig. 3.9). Cross spectra are invaluable for exploring complex relationships between several different compositional parameters that each relate to different aspects of the sedimentary environment (e.g. Clemens *et al.*, 1991).

Ancient sedimentary cycles and dating

Estimating the periods of regular sedimentary cycles

Estimating the period of regular sedimentary cycles is problematic given that: (i) ancient time series are often too short to relate directly to tie points, such as stage boundaries, which have been chronostratigraphically dated, (ii) chronostratigraphic dates in the pre-Pliocene have large errors relative to the time-spans of the time series, and (iii) hiatuses and missing sedimentary cycles are difficult to detect, particularly in fine-grained sequences.

The most widely used strategy for Mesozoic and older sections is to relate biozonal boundaries, occurring within a time series, to dated stage or period boundaries. The average time-span of the biozones is then used to estimate the average period of the regular cycles. Algeo & Wilkinson (1988) objected to this approach, particularly for shallow marine sequences, because of the likelihood of calculating periods between 20 000 and 400 000 years (since net accumulation rates for shallow sediments are constrained by the rate of subsiding continental crust). However, this is often the only method available, but in facies not deposited on shallow shelves no such bias is expected. Nevertheless, considering the errors associated with both chronostratigraphic dates and with estimating the total time-span of short time series, the inferred cycle periods can only be first-order approximations anyway.

Another objection to this form of period estimation is the problem of undetected gaps. There is a tendency for some workers to assume that no cycles are missing in their sections. Anders *et al.* (1987) compiled a large database of accumulation rates for deep-sea carbonates. Their results indicate that a section spanning one million years will *on average* be 86% complete at the 20 000-year scale — that is, seven out of fifty 20 000-year intervals could be missing. Other facies are usually even less complete (Sadler, 1981). In fine-grained deep-sea sediment there is often no obvious sedimentological or palaeontological evidence for gaps at the tens of thousands of year scale. This implies that the method of cycle-period estimation outlined above yields maximum possible periods. Clearly a more satisfactory method of dating is available in continuously laminated sections if the laminae can be shown to be annual (Anderson, 1982; Olsen, 1986).

Cycle wavelength ratios can be used as a method for relating regular sedimentary cycles to particular orbital cycles (Hays *et al.*, 1976; Schwarzacher, 1987a). This has been criticized by Algeo & Wilkinson (1988) on the basis that, if wavelength ratios are used to confirm that orbital-forcing is involved, the variety of orbital-cycle combinations can give too many possible wavelength ratios. Clearly to use this method the wavelength of regular cycles needs to be tightly constrained. Additionally, it should be noted that wavelength ratios can be altered by the presence of hiatuses (Weedon, 1989).

In several studies, bundles of five short cycles have been used to argue for the occurrence of records of precession and the short eccentricity cycle. Yet, in the Palaeozoic, the precession cycle was considerably shorter than today whereas the period of the short eccentricity cycle was unaltered, so bundles of five small wavelength cycles would not necessarily be expected (Fig. 3.2). Also Walkden & Walkden (1990) showed that a range of cycle wavelengths are able to generate bundles of five cycles. Additionally, as some stratigraphic bundles are apparently related to climatic, but not orbital, variations (Fig. 3.8), groups of small-scale cycles require careful study before orbital-forcing can be inferred from the bundling.

Despite these problems the combination of demonstrating regular cyclicity, estimating periods *of the order of* tens of thousands of years, and matching wavelength ratios for the appropriate orbital cycle periods (Fig. 3.2), can be used to make a convincing case for orbital–climatic control of sedimentation. Currently, the most forceful arguments in favour of this are the Pleistocene precedents and

the absence of other processes, known to be sufficiently regular for millions of years, that have the correct periods.

Interval dating

Since Gilbert's (1895) estimate of the duration of the late Cretaceous, the desirability of using ancient orbitally-driven sedimentary cycles for interval dating has been obvious. House (1985) suggested counting sedimentary cycles directly in order to estimate the durations of biozones. One difficulty with using the rock record directly is deciding how to separate superimposed regular cycles from each other and from noise. This can now be attempted using spectral analysis and filtering. Other difficulties lie in estimating the cycle periods themselves and in detecting gaps as just described. Schwarzacher's (1987b) comparison of time-depth plots, from counts of Cretaceous limestone–shale bundles (believed to represent 100 000 years each), with time-depth plots from magnetic reversal data demonstrated that the errors involved with interval dating may not be large, at least in some cases.

The best example of dating using proxy-palaeoclimatic time series is provided by work on Pliocene/Pleistocene oxygen isotope records (Fig. 3.3; e.g. Hays *et al.*, 1976; Imbrie *et al.*, 1984). Tuning the records is possible because a target insolation curve is available for the last 5 Ma and the ages of magnetic reversals can provide constraints. Additionally, a great many time series are available for comparison so that by pattern-matching it becomes clear whether gaps are present.

However, the results of this procedure depend on the reliability of the insolation curve and, currently, on the assumption of a constant time-lag between the changes in insolation patterns and the response of the climate (mainly ice sheets in this case). Recently it has been demonstrated that this time-lag is in fact variable in the late Pleistocene as a result of non-linearities in the climate system (Pisias *et al.*, 1990). This must have increased the dating errors of the tuning method for short time intervals. Nevertheless, the chronostratigraphic ages of magnetic reversals provides some independent indication that the tuned curves are reliable over long time intervals.

Ironically, the ages of the magnetic reversals are now under scrutiny. When Ruddiman *et al.* (1989) set up their extended standard isotope record for DSDP Site 607 they assumed that the dates were very accurate. However, with a high-resolution record from the East Pacific which accumulated more uniformly than the North Atlantic sediments, Shackleton *et al.* (1990) were able to show that the reversal dates were too young by counting cycles backwards. Support for this comes from time-depth plots for several sites. In all cases the Ruddiman *et al.* (1989) time-scale generates a kink near the Brunhes/Matuyama boundary whereas this is absent using the time-scale of Shackleton *et al.* (1990).

Independent and concurrent indications that the currently used magnetic reversal dates are too young came from the Pliocene Trubi Formation of Sicily (Hilgen & Langereis, 1988, 1989). Correlation of well-exposed sections showed that no gaps are present spanning more than about 10 000 years. Spectral analysis was used to demonstrate regular cyclicity attributed to the precession and short eccentricity cycles. Then cycle periods were initially estimated using the measured positions of magnetic reversal boundaries and published dates. This yielded periods that were consistently too small, so instead it was suggested that the reversal dates were in error. A refined set of estimates for the reversal ages has now come from matching bundles of sapropels from the Trubi marls to the calculated precession cycles from 1.5 to 3 Ma BP (Hilgen, 1991). This formation may eventually provide the best reference section for extending the standard oxygen isotope curve back to the latest Miocene (Langereis & Hilgen, 1991). Very recently, direct support for at least part of the revised time-scale of Shackleton *et al.* (1990) and Hilgen (1991) has come from single crystal Ar-Ar dating from the classic Olduvai Gorge section (Walter *et al.*, 1991).

Herbert & Hondt (1990) studied deep-sea cores using spectral analysis of colour densitometry records from the South Atlantic that span the Cretaceous/Tertiary boundary. Counts of numbers of cycles attributed to orbital precession showed that the Cretaceous/Tertiary boundary occurred about 350 000 years after the start of magnetic Chron C29R. The Chron as a whole was estimated as lasting between 515 000 and 580 000 years, in agreement with previous dating. A step-like halving of accumulation rates across the South Atlantic at the Cretaceous/Tertiary boundary was also shown.

Gale (1989) studying the British Lower Chalk, correlated many measured sections and argued that since local gaps could be identified, the sections containing the same maximum numbers of couplets are complete at the couplet scale. Bundles with an average of five couplets were ascribed to the eccentricity cycle and numbers of bundles were used to show that the duration of the Cenomanian was 4.4 m.y. However, no spectral analysis was used to demonstrate that the bundles are regular although, in view of the irregular Belemnite Marls bundles, this now appears to be a necessary step (Weedon & Jenkyns, 1990). Additionally, there was no comparison of interval durations estimated using bundles with those calculated based on couplets. Finally, Gale (1989) showed that sequence boundaries marked by intense bioturbation occur in even the most complete sections. Thus it is possible that the discrepancy between his and De Boer's (1982) estimate of the time-span of the Italian Cenomanian (7.0 m.y.) might be accounted for by basin-wide gaps at the sequence boundaries in the shallower-water British sections.

These examples demonstrate that interval dating is feasible when the period of regular cyclicity can be estimated reliably, and when many sections can be correlated to check for gaps. The existence of more than one order of regular cycle can be used to provide an internal check on any dating. However, from the last case it appears that, even when based on time series from many sections, this method can only provide a minimum estimate of interval duration. Note that an additional benefit of using multiple time series is that both lateral and temporal variations in net accumulation rates can be examined. Long time series will also permit bed-scale variations in composition to be related to sequence stratigraphic investigations.

Conclusions

Taken together, the Pleistocene palaeoceanographic precedents, the regularity and periods of the sedimentary cyclicity and studies of the likely environmental changes involved, make a strong case for there being many examples of orbital–climatic control of sedimentary cyclicity in a wide range of facies since the Proterozoic. Indeed there is even evidence for similar cyclicity in the polar layered-deposits of Mars (Cutts & Lewis, 1982).

A key contribution to sedimentology and stratigraphy made by workers on orbital–climatic cyclicity is the high stratigraphic resolution with which sections are examined. Another key contribution and indeed the backbone to such studies is the production of time series of changes in the depositional environment. Such records must be produced with due regard for diagenetic masking and aliasing. The time series can then be analysed by using: power spectra to detect regular cyclicity; filtering to isolate individual cycles; cross spectra to compare pairs of records; and evolutionary or sliding-window spectra to trace rapid variations in mean sedimentation rates. There is no shortage of environmental/sedimentological models to explain the cyclicity in terms of regional climatic processes or glacioeustatic control of relative sea-level. Yet choosing the correct model in each case remains problematic. Directly estimating sedimentary cycle periods is difficult prior to the Pliocene, but indirect methods are available.

Partly because it is non-linear the climate system possesses a red-noise spectrum with the regular orbital-forcing distinguishable as spectral peaks. Thus background climatic variations increase in amplitude for larger and larger time periods. This means that climate-related *irregular* sedimentary cycles with average periods which span the Milankovitch band of frequencies can be expected (Weedon & Jenkyns, 1990). Few studies have yet been directed to consideration of highly non-linear behaviour of the climate, but it should be borne in mind that chaotic systems can generate regular cyclicity for short periods (Stewart, 1990).

Soon, another key contribution of the field will probably be interval dating. This requires an attempt to identify local and regional stratigraphic gaps — which to some extent can be achieved using time series from several places. Recently, various methods of rapid time-series generation and analysis have been adopted which make this feasible. Interval dating could allow estimates of biozone durations and thus information on rates of evolution. And from the view-point of basin analysis, interval dating could also allow investigations of both lateral and temporal variations in net accumulation rate — which in some cases might be related to the sequence stratigraphy.

Clearly then, the field has now acquired the potential to make significant contributions to several different areas of geology.

Acknowledgements

Nick Shackleton (Cambridge) kindly supplied the isotope data for ODP Site 677 in Figs 3.3 and 3.4 and the time-scale of Shackleton *et al.* (1990). Andre Berger and M.F. Loutre (Leuven) provided advice concerning Fig. 3.1 and Matt Lloyd (Cambridge) helped draft it. Thanks also to Nick McCave (Cambridge), Tim Astin (Reading) and an anonymous reviewer for their critisisms of the manuscript. This work was supported by a Research Fellowship from Downing College, Cambridge.

References

Adéhmar, J.A. (1842) *Révolution de la mer*. Privately published: Paris.

Algeo, T.J. & Wilkinson, B.H. (1988) Periodicity of mesoscale Phanerozoic sedimentary cycles and the role of Milankovitch orbital modulation. *J. Geol.* **96**, 313–322.

Anders, M.N., Krueger, S.W. & Sadler, P.M. (1987) A new look at sedimentation rates and the completeness of the stratigraphic record. *J. Geol.* **95**, 1–14.

Anderson, R.Y. (1982) A long geoclimatic record from the Permian. *J. Geophys. Res.* **87C**, 7285–7294.

Anderson, R.Y. (1984) Orbital forcing of evaporite sedimentation. In: *Milankovitch and Climate* (Eds A. Berger, J. Imbrie, J. Hays, G. Kukla & B. Saltzman) pp. 147–162. Reidel: Dordrecht.

Arthur, M.A., Dean, W.E., Bottjer, D. & Scholle, P.A. (1984) Rhythmic bedding in Mesozoic–Cenozoic pelagic carbonate sequences: the primary and diagenetic origin of Milankovitch-like cycles. In: *Milankovitch and Climate* (Eds A. Berger, J. Imbrie, J. Hays, G. Kukla & B. Saltzman) pp. 191–222. Reidel: Dordrecht.

Barron, E.J., Arthur, M.A. & Kauffman, E.G. (1985) Cretaceous rhythmic bedding sequences: a plausible link between orbital variations and climate. *Earth Planet. Sci. Lett.* **72**, 139–172.

Berger, A. (1984) Accuracy and frequency stability of the Earth's orbital elements during the Quaternary. In: *Milankovitch and Climate* (Eds A. Berger, J. Imbrie, J. Hays, G. Kukla & B. Saltzman) pp. 3–39. Reidel: Dordrecht.

Berger, A. (1988) Milankovitch Theory and climate. *Rev. Geophys.* **26**, 624–657.

Berger, A. (1989) The spectral characteristics of pre-Quaternary climate records, an example of the relationship between Astronomical Theory and Geosciences. In: *Climate and Geosciences* (Eds A. Berger, S. Schneider & J.C. Duplessy) pp. 47–76. Kluwer Academic Publishers: The Netherlands.

Berger, A., Loutre, M.F. & Dehant V. (1989) Influence of the changing lunar orbit on the astronomical frequencies of pre-Quaternary insolation patterns. *Paleoceanogr.* **4**, 555–564.

Berger, A., Loutre, M.F. & Laskar, J. (1991) *Pre-Quaternary Astronomical Frequencies*. Scientific Report 1991/2, Institut d'Astronomie et de Geophysique G. Lemaitre, Universite Catholique de Louvain, Louvain-la-Neuve.

Cisne, J.L. (1986) Earthquakes recorded stratigraphically on carbonate platforms. *Nature* **323**, 320–322.

Clemens, S.C., Prell, W.L., Murray, D., Shimmfield, G.B.

& Weedon, G.P. (1991) Forcing mechanisms of the Indian Ocean Monsoon. *Nature* **353**, 720–725.

Clifton, H.E. (1981) Progradational sequences in Miocene shoreline deposits, southeastern Caliente Range, California. *J. Sedim. Petrol.* **51**, 165–184.

Cottle, R.A. (1989) Orbitally mediated cycles from the Turonian of southern England: their potential for high-resolution stratigraphic correlation. *Terra Nova* **1**, 426–432.

Croll, J. (1864) On the physical cause of the change of climate during geological epochs. *Phil. Mag.* **28**, 121–137.

Cutts, J.A. & Lewis, B.H. (1982) Models of climate cycles recorded in Martian polar layered deposits. *Icarus* **50**, 216–244.

De Boer, P.L. (1982) Cyclicity and the storage of organic matter in Middle Cretaceous pelagic sediments. In: *Cyclic and Event Stratification* (Eds G. Einsele & A. Seilacher) pp. 456–475. Springer-Verlag: Berlin.

Ditchfield, P. & Marshall, J.D. (1989) Isotopic variation in rhythmically bedded chalks: palaeotemperature variation in the Upper Cretaceous. *Geology* **17**, 842–845.

Eicher, D.L. & Diner, R. (1989) Origin of the Cretaceous Bridge Creek cycles in the Western Interior, United States. *Palaeogeogr. Palaeoclim. Palaeoecol.* **74**, 127–146.

Einsele, G. (1982) Limestone–marl cycles (periodites): diagnosis, significance, causes — a review. In: *Cyclic and Event Stratification* (Eds G. Einsele & A. Seilacher) pp. 8–53. Springer-Verlag: Berlin.

Emiliani, C. (1955) Pleistocene temperatures. *J. Geol.* **63**, 538–578.

Emiliani, C. (1966) Palaeotemperature analysis of Carribean cores P6304-8 and P6304-9 and a generalized temperature curve for the past 425 000 years. *J. Geol.* **74**, 109–126.

Erickson, M.C. & Slingerland, R. (1990) Numerical simulation of tidal and wind-driven circulation in the Cretaceous Interior Seaway of North America. *Geol. Soc. Am. Bull.* **102**, 1499–1516.

Fairbanks, R.G. (1989) A 17 000 year glacio-eustatic sea level record: influence of glacial melting rates on the Younger Dryas event and deep-ocean circulation. *Nature* **342**, 637–642.

Farrell, J.W. & Prell, W.L. (1989) Climatic change and $CaCO_3$ preservation: an 800 000 year bathymetric reconstruction from the central equatorial Pacific Ocean. *Paleoceanogr.* **4**, 447–466.

Fischer, A.G. (1980) Gilbert — bedding rhythms and geochronology. *Special Pap. Geol. Soc. Am.* **183**, 93–104.

Fischer, A.G. (1986) Climatic rhythms recorded in strata. *Ann. Rev. Earth Planet. Sci.* **14**, 351–376.

Fischer, A.G., Herbert, T.D. & Premoli Silva I. (1985) Carbonate bedding cycles in Cretaceous pelagic and hemipelagic sequences. In: *Fine Grained Deposits and Biofacies of the Western Interior Seaway: Evidence of Cyclic Sedimentary Processes* (Eds L. Pratt & E. Kauffman). Soc. Econ. Paleont. Miner. Guidebook 4, 1–10.

Fischer, A.G., De Boer, P.L. & Premoli Silva, I. (1990) Cyclostratigraphy. In: *Cretaceous Resources, Events and Rhythms* (Eds R.N. Ginsburg & B. Beaudoin) pp. 139–172. Kluwer Academic Publishers: The Netherlands.

Foucault, A., Powichrowski, L. & Prud'Homme, A. (1987) Le Contrôle astronomique de la sédimentation turbidi-tique: example du Flysch à Helminthoides des Alpes Ligures (Italie). *Comptes Rendus Academie Sci. Paris* Series II, **305**, 1007–1011.

Frakes, L.A. (1979) *Climates Throughout Geologic Time.* Elsevier: Amsterdam.

Gale, A.S. (1989) A Milankovitch scale for Cenomanian time. *Terra Nova* **1**, 420–425.

Genthon, C., Barnola, J.M., Raynaud, D., Lorius, C., Jouzel, J., Barkov, N.I., Korotkevich, Y.S. & Kotlyakov, V.M. (1987) Vostok ice core: climatic response to CO_2 and orbital-forcing changes over the last climatic cycle. *Nature* **329**, 414–418.

Gilbert, G.K. (1895) Sedimentary measurement of geologic time. *J. Geol.* **3**, 121–127.

Glancy, T.J., Barron, E.J. & Arthur, M.A. (1986) An initial study of the sensitivity of modelled Cretaceous climate and cyclic insolation forcing. *Paleoceanogr.* **1**, 523–537.

Goldhammer, R.K., Dunn, P.A. & Hardie, L.A. (1990) Depositional cycles, composite sea-level changes, cycle stacking patterns and the hierarchy of stratigraphic forcing: examples from Alpine Triassic platform carbonates. *Geol. Soc. Am. Bull.* **102**, 535–562.

Haak, A.B. & Schlager, W. (1989) Compositional variations in calciturbidites due to sea-level fluctuations, late Quaternary, Bahamas. *Geol. Rdsch.* **78**, 477–486.

Hay, W.H., Behensey, J.F., Barron, E.J. & Sloan, J.L. (1982) Late Triassic–Liassic palaeoclimatology of the Proto-central North Atlantic rift system. *Palaeogeog. Palaeoclim. Palaeoecol.* **40**, 13–30.

Hays, J.D., Imbrie, J. & Shackleton, N. J. (1976) Variations in the Earth's orbit: pacemaker of the ice ages. *Science* **194**, 1121–1132.

Herbert, T.D. & Fischer, A.G. (1986) Milankovitch climatic origin of mid-Cretaceous black shale rhythms in central Italy. *Nature* **321**, 739–743.

Herbert, T.D. & Hondt, S.L. (1990) Precessional climate cyclicity in Late Cretaceous–Early Tertiary marine sediments: a high resolution chronometer of Cretaceous–Tertiary boundary events. *Earth Planet. Sci. Lett.* **99**, 263–275.

Herbert, T.D., Stallard, R.F. & Fischer, A.G. (1986) Anoxic events, productivity rhythms, and the orbital signature in a mid-Cretaceous deep sea sequence from central Italy. *Paleoceanogr.* **1**, 495–506.

Herschel, J.F.W. (1832) On the astronomical causes which may influence geological phenomena. *Trans. Geol. Soc.* 2nd Series, **3**, 393–399.

Hilgen, F.J. (1991) Astronomical calibration of Gauss to Matuyama sapropels in the Mediterranean and implication for the geomagnetic polarity timescale. *Earth Planet. Sci. Lett.* **104**, 226–244.

Hilgen, F.J. & Langereis, C.G. (1988) The age of the Miocene–Pliocene boundary in the Capo Rossello area (Sicily). *Earth Planet Sci. Lett.* **91**, 214–222.

Hilgen, F.J. & Langereis, C.G. (1989) Periodicities of $CaCO_3$ cycles in the Pliocene of Sicily: discrepancies with the quasi-periods of the Earth's orbital cycles? *Terra Nova* **1**, 409–415.

House, M.R. (1985) A new approach to an absolute timescale from measurements of orbital cycles and sedimentary microrhythms. *Nature* **315**, 721–725.

Imbrie, J. & Imbrie, K.P. (1979) *Ice Ages, Solving the Mystery.* Enslow: USA.

Imbrie, J. & Imbrie, J.Z. (1980) Modelling the climatic response to orbital variations. *Science* **207**, 943–953.

Imbrie, J., Hays, J.D., Martinson, D.G., McIntyre, A., Mix, A.C., Morley, J.J., Pisias, N.G., Prell, W.L. & Shackleton, N.J. (1984) The orbital theory of Pleistocene climate: support from a revised chronology of the marine $\delta^{18}O$ record. In: *Milankovitch and Climate* (Eds A. Berger, J. Imbrie, J. Hays, G. Kukla & B. Slatzman) Part 1, pp. 269–305. Reidel: Dordrecht.

James, I.N. & James, P.M. (1989) Ultra-low-frequency variability in a simple atmospheric circulation model. *Nature* **342**, 53–55.

Laferriere, A.P. & Hattin, D.E. (1989) Use of rhythmic bedding patterns for locating structural features, Niobrara Formation, United States Western Interior. *Am. Ass. Petrol. Geol. Bull.* **73**, 630–640.

Langereis, C.G. & Hilgen, F.J. (1991) The Rossello composite: a Mediterranean and global reference section for the Early to early Late Pliocene. *Earth Planet. Sci. Lett.* **104**, 211–225.

Laskar, J. (1989) A numerical experiment on the chaotic behaviour of the solar system. *Nature* **338**, 237–238.

Le Treut, H. & Ghil, M. (1983) Orbital forcing, climatic interactions and glaciation cycles. *J. Geophys. Res.* **88**, 5167–5190.

Lyell, C. (1872) *Principles of Geology*, 2 vols, 11th edn. John Murray: London.

Melnyk, D.H. & Smith, D.G. (1989) Outcrop to subsurface cycle correlation in the Milankovitch frequency band: Middle Cretaceous, central Italy. *Terra Nova* **1**, 432–436.

Milankovitch, M. (1941) *Canon of Insolation and the Ice-Age Problem* (English translation by Isreal Program for Scientific Translation, Jerusalem 1969). R. Serbian Acad. Spec. Publ. 132.

Milankovitch, V. (1984) The memory of my father. In: *Milankovitch and Climate* (Eds A. Berger, J. Imbrie, J. Hays, G. Kukla & B. Saltzman) Part 1, pp. xxii–xxxiv. Reidel: Dordrecht.

Murray, D.W. & Prell, W.L. (1991) Pliocene to Pleistocene variations in calcium carbonate, organic carbon, and opal on the Owen Ridge, North Arabian Sea. *Proc. Ocean Dril. Prog. Sci. Res.* **117**, 343–363.

Napoleone, G. & Ripepe, M. (1989) Cyclic geomagnetic changes in mid-Cretaceous rhythmites, Italy. *Terra Nova* **1**, 437–442.

Olsen, P. (1986) A 40 million year lake record of early Mesozoic orbital climatic forcing. *Science* **234**, 842–848.

Pestiaux, P. & Berger, A. (1984) An optimal approach to the spectral characterization of deep-sea climate records. In: *Milankovitch and Climate* (Eds A. Berger, J. Imbrie, G. Kukla & B. Saltzman) pp. 417–445. Reidel: Dordrecht.

Pisias, N.G. & Mix, A.C. (1988) Aliasing of the geological record and the search for long-period Milankovitch cycles. *Paleoceanogr.* **3**, 613–619.

Pisias, N.G., Mix, A.C. & Zahn, R. (1990) Nonlinear response in the global climate system: evidence from benthonic oxygen isotope record in Core RC13-110. *Paleoceanogr.* **5**, 147–160.

Prell, W.L., Niitsuma, N., Emeis, K.C. *et al.* (1989) *Proc Ocean Dril. Prog. Init. Rep.* **117**.

Priestley, M.B. (1981) *Spectral Analysis and Time Series*, Vols 1 and 2. Academic Press: London.

Raymo, M.E., Ruddiman, W.F., Backman, J., Clement, B.M. & Martinson, D.G. (1989) Late Pliocene variations in northern hemisphere ice sheets and North Atlantic deep-water circulation. *Paleoceanogr.* **4**, 413–446.

Read, J.F. & Goldhammer, R.K. (1988) Use of Fischer plots to define third-order sea-level curves in Ordovician peritidal cyclic carbonates, Appalachians. *Geology* **16**, 895–899.

Read, J.F., Grotzinger, J.P., Bova, J.A. & Koerschner, W.F. (1986) Models for generation of carbonate cycles. *Geology* **14**, 107–110.

Rossignol-Strick, M. (1983) African Monsoons, an immediate climate response to orbital insolation. *Nature* **304**, 46–49.

Rossignol-Strick, M., Nesteroff, W., Olive, P. & Vergnaud-Grazzini, C. (1982) After the deluge: Mediterranean stagnation and sapropel formation. *Nature* **295**, 105–110.

Ruddiman, W.F. (1985) Climate studies in ocean cores. In: *Paleoclimate Analysis and Modelling* (Ed. A.D. Hecht) pp. 197–257. Kluwer Academic Publishers: The Netherlands.

Ruddiman, W.F. & McIntyre, A. (1981) Oceanic mechanisms for amplification of the 23 000-year ice-volume cycle. *Science* **212**, 617–627.

Ruddiman, W.F., Raymo, M.E., Martinson, D.G., Clement, B.M. & Backman, J. (1989) Pleistocene evolution: northern hemisphere ice sheets and North Atlantic Ocean. *Paleoceanogr.* **4**, 353–412.

Sadler, P.M. (1981) Sediment accumulation rates and the completeness of stratigraphic sections. *J. Geol.* **89**, 569–584.

Saltzman, B. & Maasch, K.A. (1990) A first-order global model of late Cenozoic climate change. *Trans. R. Soc. Edinb.: Earth Sci.* **81**, 315–325.

Schuster, H.G. (1984) *Deterministic Chaos: An Introduction.* Weinheim: Germany.

Schwarzacher, W. (1964) An application of statistical time series analysis to a limestone–shale sequence. *J. Geol.* **72**, 195–213.

Schwarzacher, W. (1975) *Sedimentation Models and Quantitative Stratigraphy.* Developments in Sedimentology 19. Elsevier: Amsterdam.

Schwarzacher, W. (1987a) The analysis and interpretation of stratification cycles. *Paleoceanogr.* **2**, 79–95.

Schwarzacher, W. (1987b) Astronomically controlled cycles in the lower Tertiary of Gubbio (Italy). *Earth Planet. Sci. Lett.* **84**, 22–26.

Schwarzacher, W. (1989) Milankovitch-type cycles in the Lower Carboniferous of NW Ireland. *Terra Nova* **1**, 468–473.

Schwarzacher, W. & Haas, J. (1986) Comparative statistical analysis of some Hungarian and Austrian Upper Triassic peritidal carbonate sequences. *Acta Geol. Hung.* **29**, 179–196.

Shackleton, N.J. & Imbrie, J. (1990) The $\delta^{18}O$ spectrum of oceanic deep water over a five-decade band. *Clim. Change* **16**, 217–230.

Shackleton, N.J. & Opdyke, N.D. (1973) Oxygen isotope and palaeomagnetic stratigraphy of equatorial Pacific core V28-238: oxygen isotope temperatures and ice volumes on a 10^5 year and 10^6 year scale. *Quat. Res.* **3**, 39–55.

Shackleton, N.J. & Pisias, N.G. (1985) Atmospheric carbon dioxide, orbital forcing and climate. In: *The Carbon Cycle and Atmospheric Co₂: Natural Variations Archean to Present* (Eds E.T. Sundquist & W.S. Broecker) Am. Geophys. Un., Geophys. Monogr. **32**, 303–317.

Shackleton, N.J., Berger, A.L. & Peltier, W.R. (1990) An alternative astronomical calibration of the Lower Pleistocene timescales based on ODP Site 677. *Trans. R. Soc. Edinb.: Earth Sci.* **81**, 251–261.

Spicer, R.A. & Parrish J.T. (1990) Late Cretaceous–early Tertiary palaeoclimates of northern high latitudes: a quantitative view. *J. Geol. Soc. Lond.* **147**, 329–341.

Stewart, I. (1990) *Does God Play Dice? – The New Mathematics of Chaos.* Penguin: London.

Van Echelpoel, E. & Weedon, G.P. (1990) Milankovitch cyclicity and the Boom Clay Formation: an Oligocene siliciclastic shelf sequence in Belgium. *Geol. Mag.* **127**, 599–604.

Van Tassell, J. (1987) Upper Devonian Catskill delta margin cyclic sedimentation: Brallier, Scherr and Foreknobs Formations of Virginia and West Virginia. *Geol. Soc. Am. Bull.* **99**, 414–426.

Walkden, G.M. & Walkden, G.D. (1990) Cyclic sedimentation in carbonate and mixed carbonate–clastic environments: four simulation programs for a desktop computer. *Spec. Publ. Int. Ass. Sediment.* **9**, 55–78.

Walter, R.C., Manega, P.C., Hay, R.L., Drake, R.E. & Curtis, G.H. (1991) Laser-fusion ^{40}Ar-^{39}Ar dating of Bed 1 Olduvai Gorge, Tanzania. *Nature* **354**, 145–149.

Watkins, D.K. (1989) Nannoplankton productivity fluctuations and rhythmically-bedded pelagic carbonates of the Greenhorn Limestone (Upper Cretaceous). *Palaeogeogr. Palaeoclim. Palaeoecol.* **74**, 75–86.

Weedon, G.P. (1985) Hemipelagic shelf sedimentation and climatic cycles: the basal Jurassic (Blue Lias) of South Britain. *Earth Planet. Sci. Lett.* **76**, 321–335.

Weedon, G.P. (1989) The detection and illustration of regular sedimentary cycles using Walsh power spectra and filtering, with examples from the Lias of Switzerland. *J. Geol. Soc. Lond.* **146**, 133–144.

Weedon, G.P. (1991) The spectral analysis of stratigraphic time series. In: *Cycles and Events in Stratigraphy* (Eds G. Einsele, W. Ricken & A. Seilacher) pp. 840–854. Springer-Verlag: Berlin.

Weedon, G.P. & Jenkyns, H.C. (1990) Regular and irregular climatic cycles and the Belemnite Marls (Pliensbachian, Lower Jurassic, Wessex Basin). *J. Geol. Soc. Lond.* **147**, 915–918.

Weedon, G.P., Robinson, S.G. & Jenkyns, H.C. (in preparation) Magnetic susceptibility as a high-resolution logging tool for Mesozoic mudrocks.

Williams, G.E. (1991) Milankovitch-band cyclicity in bedded halite deposits contemporaneous with Late Ordovician–Early Silurian glaciation, Canning Basin, Western Australia. *Earth Planet. Sci. Lett.* **103**, 143–155.

4   Carbonate diagenesis and sequence stratigraphy

MAURICE E. TUCKER

Introduction

The diagenesis of carbonate sediments has been a major topic of research for many decades, with fundamental observations being made by Sorby back in the mid-19th century. Quaternary carbonate sediments of many regions have now been studied intensively for the effects of diagenesis, notably in surface and shallow-burial marine and meteoric diagenetic environments. The early (marine and meteoric) and late (burial) diagenesis of many ancient limestones has also been explored in innumerable case histories, with comparisons frequently made to the Recent. Thus there is an extensive literature on carbonate diagenesis, and this has been reviewed recently in Moore (1989), McIlreath & Morrow (1990), Tucker & Bathurst (1990), Tucker & Wright (1990) and Tucker (1991a). However, there has been little attempt to synthesize the information to produce *diagenetic models*, in the same way that facies models were developed in the 1960s and 1970s as a distillation product of all the sedimentological data that had been accumulated. Diagenetic models should be useful for understanding and predicting the paths of carbonate diagenesis, and also for the prediction of porosity, one of the main factors in hydrocarbon reservoir potential.

With the development of sequence stratigraphy in the last decade and its more widespread application in the last few years, it is possible to integrate carbonate diagenesis into the patterns of relative sea-level change, which are the underlying control on the formation of sequences and their systems tracts. Two other factors of great importance to the type of diagenesis affecting a limestone are the prevailing climate and the carbonate sediment mineralogy. The latter has a major influence on diagenetic potential and is controlled by seawater chemistry, environmental conditions and skeletal evolution/extinction. Climate, seawater chemistry, carbonate sediment mineralogy and even the nature of sequences themselves and the patterns of relative sea-level change, have all varied through geological time, so that, as a consequence, carbonate diagenesis has also varied through time.

Putting carbonate diagenesis into a sequence stratigraphic framework facilitates a better understanding of diagenetic patterns in a limestone succession and permits a degree of prediction. This paper explores the relationship between carbonate diagenesis and relative sea-level change through the Phanerozoic.

Carbonate diagenesis

Three major carbonate diagenetic environments are distinguished: marine, meteoric and burial (see Fig. 4.1), with the first two being predominantly surface-related and shallow-burial environments. Diagenetic environments are briefly discussed here in terms of major controls and processes, and porosity loss and gain (for reviews see McIlreath & Morrow, 1990; Tucker & Wright, 1990 and Tucker, 1991a). Dolomitization is, of course, also an important process in many limestones and is still a topic of

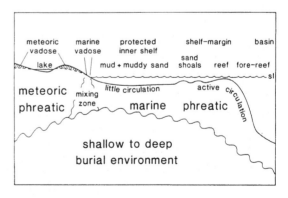

Fig. 4.1 Carbonate diagenetic environments, schematically drawn for a rimmed shelf with unconfined aquifers (from Tucker, 1991a).

great debate. The process operates in the three diagenetic environments and it is treated separately here. Dolomite dissolution and gypsum–anhydrite–halite cementation and dissolution are additional, but often neglected processes in carbonate successions, operating at depth and near-surface, which have great significance for reservoir potential.

Marine diagenesis

The main change in porosity during marine diagenesis is one of porosity loss through cementation (see Fig. 4.2a,b). Cements are commonly precipitated in intragranular and intergranular cavities, and porosity can be lost completely (Fig. 4.2b). The cements are mostly precipitated directly from seawater, and many thousands of pore volumes of water must pass through a pore to occlude the porosity. Microbial processes may be important in the precipitation of some types of marine cement. Internal sedimentation in cavities, which are common in reef-rocks, is also important in porosity

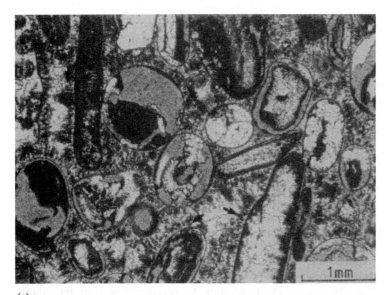

(a)

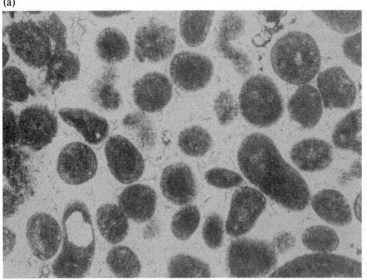

(b)

Fig. 4.2 (a) Oolitic-skeletal grainstone cemented with early isopachous acicular calcite, interpreted as originally high-Mg calcite of marine origin. There is some later calcite spar cement. The ooids, originally of aragonite, have been partially to completely dissolved out; the porosity is shown by a grey colour in the photomicrograph (blue resin). The marine cement was also subjected to leaching, seen by the impregnation with stained resin (appearing dark here, see arrows). This limestone was first cemented on the sea-floor and then on exposure subjected to meteoric dissolution of aragonitic grains and high-Mg calcite cement, with minor calcite spar precipitation. (b) Calcitic ooids with moderately preserved radial structure cemented by isopachous marine calcite cement. The early cementation prevented later burial compaction. Jurassic, Smackover Formation, subsurface northern Louisiana.

reduction, especially if the internal sediments are subsequently cemented. The dominant controls on marine cementation are the degree of seawater circulation through the sediments, the climate and the sedimentation rate.

In modern environments, marine cementation is most common along windward platform margins where vast quantities of seawater are pumped through the sediments by waves, storms and tidal currents. The water agitation results in CO_2-degassing and this promotes $CaCO_3$ precipitation. There may also be a warming of the seawater as the open-ocean water comes into the shallows of the carbonate platform; this too promotes precipitation. Reefs and oolite shoals of rimmed shelves commonly contain abundant marine cements, and these can lead to the development of steep fore-reef slopes and hardgrounds. With extensive sea-floor cementation of grainstones, tepees may form, with sheet cracks, internal sediments and cement botryoids. Hardgrounds are important in terms of fluid migration, commonly acting as permeability barriers. Leeward margins and shelf lagoons, by way of contrast, generally show less marine cementation, largely because porewaters are moving very slowly through the sediments in this more stagnant marine diagenetic environment.

Climate is important in the cementation of shallow-marine sediments. Where there is a high degree of aridity, then seawater on a carbonate platform is subject to evaporation and this will promote the precipitation of cements on the shallow sea-floor. The rate of carbonate production will also affect the degree of sea-floor cementation. Where production rates are low, there is more likelihood of cementation. Hardgrounds are best developed in areas of slow sedimentation. Thus, generalizing, marine cementation is likely to be most extensive on windward margins in arid regions where sedimentation rates are low.

Below several metres within shallow-marine carbonate sediments, the rate of pore-fluid movement is drastically reduced and little cementation can take place. Microbial processes may operate, such as micritization, which can lead to the development of grain microporosity. However, there is usually no further porosity loss in a carbonate sediment buried in marine porewaters at shallow to moderate depths, until burial compaction–cementation processes take over at several hundred metres burial depth.

Marine diagenesis also operates along shorelines, notably in arid regions, in the intertidal and supratidal zones. Cements are precipitated in sands to form beachrocks and in tidal-flat muds to give crusts, tepees and pisoids. Dolomitization from the evaporative concentration of seawater is also common on supratidal flats, and gypsum precipitation may be associated. These surface-related processes, mostly a product of evaporation and flood recharge of the supratidal zone, constitute the marine subaerial diagenetic environment.

Modern cements in tropical shallow-marine sediments are composed of aragonite and high-Mg calcite (11–19 mole % $MgCO_3$), typically in acicular and botryoidal, and bladed and peloidal fabrics respectively. Where these cements were precipitated in ancient limestones, they have mostly since been replaced by calcite. However, it is now clear that at times in the past, calcite (with a low to moderate Mg content) was a marine cement. The fabric of the calcite varies from fibrous crystals (Fig. 4.2a), which may be of the radiaxial type, to equant crystals. Syntaxial calcite overgrowths upon echinoderm debris, an important component of Palaeozoic and Mesozoic shallow-water carbonates, were also precipitated on the sea-floor, although this does not appear to happen now. The geological record of marine cements is discussed again later in this paper (see Fig. 4.13) and also reviewed in Tucker & Wright (1990) and Tucker (1992).

At the present time, dissolution of grains on the sea-floor does take place in relatively deep water (more than several hundred metres) and in higher latitudes. Modern low-latitude, shallow-water carbonate sediments, the equivalent of the majority of ancient limestones, are generally not subject to any dissolution on the sea-floor. However, it does appear that, in times past, shallow tropical seas were undersaturated with respect to aragonite, so that dissolution of aragonitic grains did take place. These were times when calcite was the dominant marine precipitate, in the mid-Palaeozoic and Jurassic/Cretaceous (e.g. Palmer *et al.*, 1988). Thus it is possible that some dissolutional porosity was generated on the sea-floor in some ancient limestones.

Meteoric diagenesis

In the near-surface meteoric diagenetic environment, porosity can be gained or lost. Porosity reduction mostly occurs through cementation (plus internal sedimentation) and this may take place in the vadose and/or phreatic zone (Fig. 4.3). The cement is usually calcite spar, as a result of the low molar (0.4) Mg : Ca ratio of meteoric waters. Cementation varies considerably in extent, from simply occurring at grain contacts in the vadose zone, to total cementation in the uppermost vadose zone, where pedogenic processes are involved. Cementation also takes place in the lower vadose/phreatic zone, where downward-percolating waters have become supersaturated with respect to CaCO$_3$ and partial to complete lithification occurs (e.g. Fig. 4.3b).

Porosity can be created in carbonate sediments/limestones in the meteoric environment through leaching of grains by carbonate-undersaturated waters (e.g. Figs 4.2a and 4.3a) and through karstification, the wholesale dissolution of limestone to form vugs and caverns.

There are several important controls on the degree of meteoric diagenesis: climate, amplitude and duration of sea-level fluctuations causing subaerial exposure, and original sediment mineralogy.

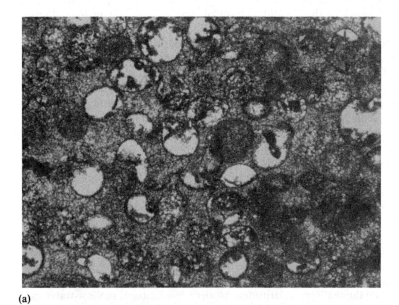

(a)

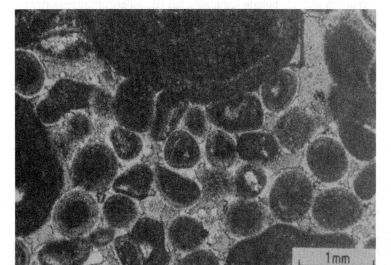

(b)

Fig. 4.3 (a) Formerly aragonitic ooids showing early, pre-compaction dissolution with some oomolds filled by post-compaction calcite spar and others empty (white in photomicrograph). The compaction distorted the oomolds. Jurassic, Smackover Formation, subsurface southern Arkansas. (b) Calcitic ooids with patchy circumgranular calcite cement precipitated in the near-surface meteoric phreatic zone. Some compaction took place as shown by the concavo-convex and sutured grain contacts. There was no later calcite spar precipitation. The intergranular grey colour is a blue resin. Jurassic, Smackover Formation, subsurface northern Louisiana.

Climate is a fundamental control on meteoric diagenesis, since the quantity and frequency of meteoric water passing through the sediments control the degree of leaching and cementation. Obviously, under a humid climate, the effects of meteoric diagenesis will be more marked. Grain leaching and karstification are major processes and at certain horizons cementation will be very extensive. The duration of the meteoric diagenetic environment is a major factor in the development of karst, with the order of one million years required for a mature karst profile to develop.

The magnitude of relative sea-level fall is important in controlling the depth to which meteoric processes operate. Karst can develop to depths of several hundred metres at a time of extreme sea-level lowstand or tectonic uplift. The presence of confined aquifers is also important, since then meteoric waters can penetrate deep into a carbonate platform and even emerge upon the sea-floor.

The original sediment mineralogy is important in terms of the degree of leaching and cementation that can take place, that is, the sediment's diagenetic potential. Modern shallow-water carbonate sediments are composed of a mixture of aragonite, high-Mg calcite and low-Mg calcite grains. Aragonite is the least stable in meteoric water and is readily dissolved (as in Figs 4.2a and 4.3a). High-Mg calcite tends to lose its magnesium to give low-Mg calcite; if the waters are very undersaturated with respect to $CaCO_3$ then the calcite will dissolve too. Thus in a mixed-mineralogy or aragonite-dominated sediment, there is a potential for porosity-gain from leaching, as well as a ready source of $CaCO_3$ for cementation, generally lower in the profile. In an aragonite-free sediment, there is no readily dissolvable carbonate, and so more undersaturated waters are required to dissolve the calcite for cementation. Dominantly calcitic sediments are likely to show less meteoric leaching and less cementation compared to aragonite-dominated ones.

Diagenetic potential is important when considering the geological record of carbonate sediments. For example, Jurassic/Cretaceous and mid-Palaeozoic oolites were dominantly composed of calcitic ooids (as those in Figs 4.2b and 4.3b), and Palaeozoic bioclastic sediments were largely calcitic too, since this was the mineralogy of the most important limestone-forming organisms of the time. Thus these sediments would have had a lower potential for early meteoric cementation compared to Mesozoic/Cenozoic bioclastic sediments, which would have contained many aragonitic bioclasts, and Permian/Triassic and Cenozoic oolites, where aragonite was the dominant ooid mineralogy. Temperate (cold-water) carbonate sediments are also dominantly calcitic and so have a lower diagenetic potential than most low-latitude tropical carbonates.

Burial diagenesis

Most burial diagenetic processes lead to a destruction of porosity. Cementation in the burial environment is by calcite spar, and in many limestones this has completely occluded the porosity. In many carbonate reservoirs, it appears that oil entry took place early, before burial spar cementation in adjacent rocks. In other cases, burial cements are present in the reservoir (e.g. Fig. 4.4a). The replacement of porewaters by hydrocarbons generally prevents any further diagenetic reactions since they can only take place in the medium of water.

The other major burial process, compaction, also results in porosity loss. Closer packing and grain fracture in mechanical compaction, and pressure dissolution between grains and along more argillaceous laminae in chemical compaction, both lower the primary porosity (see Fig. 4.4a,b). Early dissolutional porosity may also be reduced through compaction (e.g. Fig. 4.3a). Pressure dissolution is also significant in releasing $CaCO_3$ into the pore-fluids for cementation close-by or higher in the succession.

The onset of compaction in a sediment, generally at several hundred metres burial depth, induces the movement of pore-fluids. They may move vertically upwards if the sediment is homogeneous and porous, or laterally, up dip, along porous horizons if there are impermeable units within the succession. The migration of burial fluids heralds the beginning of cementation by calcite spar. There has thus been an extended period of little diagenetic alteration after the near-surface processes down to the burial depth where the overburden pressure is sufficient to induce compaction. Where there are interbedded mudrocks to de-water and provide a source of magnesium, dolomitization of limestones may occur.

Porosity may be gained during burial diagenesis through dissolution of metastable grains and early cements and compactional/tectonic fracturing of early-cemented, generally fine-grained and brittle

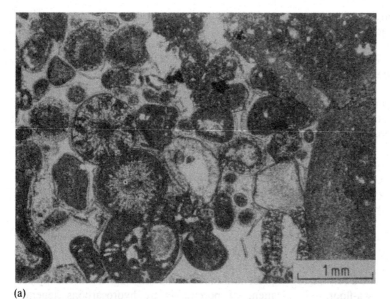

(a)

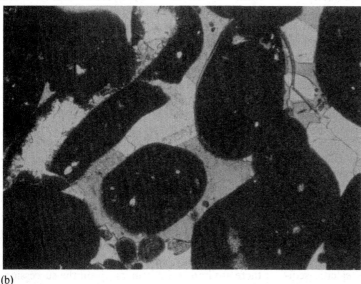

(b)

Fig. 4.4 (a) Calcitic ooids and micritized grains with very thin coats of marine cement. Mechanical compaction has led to the spalling of this cement and chemical compaction has led to sutured and concavo-convex grain contacts. Jurassic, Smackover Formation, subsurface northern Louisiana. (b) An absence of early cement in this peloidal-oolitic grainstone has led to spalling of outer oolitic lamellae and fracturing of coated bioclasts during burial. The porosity has been partially occluded by calcite spar and dolomite rhombs. Jurassic, Smackover Formation, subsurface northern Louisiana.

carbonates. The significance of burial dissolution is still a matter of some debate, but it is likely to occur in association with the maturation of organic matter and oil generation, and in the vicinity of the oil–water contact, where diagenetic reactions are likely to proceed rapidly.

The dominant controls on the burial diagenesis of carbonate sediments are burial depth and overburden pressure, pore-fluid composition and pressure, and organic maturation. However, the early diagenesis of a limestone is also important since this can determine the path of burial diagenesis. For example, a moderate amount of marine or meteoric cementation can reduce the amount of grain–grain compaction that a limestone suffers later during burial (see Figs 4.2b and 4.3a). Without the early cement, grain–grain interpenetration and pressure dissolution can be extensive; this will liberate $CaCO_3$ for cementation to reduce the porosity still further. Reservoirs in Jurassic oolites of the Paris Basin occur where marine cements retarded burial compaction, but were not so extensive so as to destroy the reservoir quality completely (Purser, 1978).

Dolomitization

Many limestones have been subjected to dolomitization and, associated with this, there may be an increase in porosity from net calcite dissolution. Various models have been put forward to account for dolomitization, but there is still much debate over the efficacy of each (see review in Tucker & Wright, 1990). The popular models are: peritidal/evaporative; reflux; mixing-zone; seawater circulation; and burial/hydrothermal. The first four are penecontemporaneous or relatively early diagenetic, near-surface processes, whereas the last is a later diagenetic process, commonly operating after the mixed mineralogy of the original sediment has stabilized to calcite. Of the models for early dolomitization, the most contentious has been the meteoric-marine mixing-zone scenario. The prevailing view now is that although in terms of thermodynamics it is a powerful mechanism, kinetic factors rule it out as a geologically significant process. It is now widely believed that most pervasively dolomitized limestones were formed in the near surface but that they were subjected to recrystallization during burial, and along with this their chemistry was reset to a greater or lesser extent. Further dolomite precipitation, as cement, also takes place during burial.

In all of the models for dolomitization, most of the magnesium for the dolomite is derived from seawater, although there may have been some modification of the chemistry. Factors thought to promote dolomitization are: lowering of seawater sulphate content; dilution of seawater (lowers ionic strength but maintains the molar Mg : Ca ratio around 5); evaporation of seawater (raises molar Mg : Ca ratio); and elevation of temperature (as during burial, overcomes some kinetic factors).

For the development of pervasive dolomites by any of the early diagenetic models, one of the main factors is the efficient pumping of the dolomitizing pore-fluids through the carbonate sediments. Such transport paths are more likely to be established during periods of relative sea-level change. Hence, the various dolomitization models can be tied into the concepts of sequence stratigraphy (see later section). However, if the rates of sea-level change are too rapid, then there may not be sufficient time for dolomitization to take place.

Several of the dolomitization models do commonly relate to sequence boundaries, the exposure horizons which delimit sequences in the concepts of sequence stratigraphy and which are the result of relative sea-level falls (see next section). Where the sequence consists of several or many parasequences, then dolomitization may relate to the parasequence boundaries, horizons representing shorter periods of exposure (see later section). Climate is another important factor in the models, with an arid climate necessary for the supratidal/evaporative and reflux models and a humid climate more likely to promote mixing-zone-related dolomitization. Thus, for early diagenetic, near-surface dolomitization, there are three main scenarios: (i) dolomitization during relative sea-level falls and under a humid climate taking place in the mixing-zone *and/or* within the zone of circulating marine water ahead of the mixing zone, (ii) supratidal/evaporative and reflux dolomitization by marine water during stillstands or relative sea-level falls and under an arid climate, and (iii) dolomitization by circulating seawater, especially during relative sea-level rises (see Fig. 4.5 and discussion in later sections).

Dolomite, and especially the calcite in partially dolomitized limestones, may be subjected to dissolution during burial or on later uplift, so creating porosity, and dolomite may also be calcitized (de-dolomitization). Dolomite dissolution during burial appears to be an important process responsible for the porosity in some hydrocarbon reservoirs (such as in the Zechstein of The Netherlands, e.g. Van der Baan, 1990, and in the Devonian of Western Canada). The dissolution may be related to the generation of CO_2-rich pore-fluids during organic maturation. De-dolomitization on uplift is commonly associated with evaporite dissolution.

Sequence stratigraphy

Sedimentary successions are commonly organized into unconformity-bound sequences, generally in the range of 50–200 m thick, and mostly produced by third-order (1–10 m.y., Table 4.1) relative sea-level changes (see Van Wagoner *et al.*, 1988, 1990 for details). Within the sequences, various systems tracts can be distinguished, deposited during specific parts of the relative sea-level change curve; these are the lowstand (LST), transgressive (TST), highstand (HST) systems tracts and shelf-margin wedge (SMW) (Fig. 4.6). Within many carbonate sequences, there are commonly metre-scale

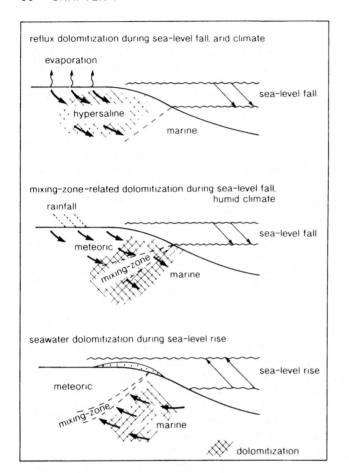

Fig. 4.5 Models for dolomitization induced by relative changes in sea-level. Mixing-zone-related dolomitization refers to dolomitization taking place within the mixing-zone (which probably does not happen) and to dolomitization taking place within the circulating marine groundwaters ahead of the mixing zone.

Table 4.1 Orders of sea-level change and mechanisms

Orders	Years	Tectonoeustatic	Rifting and thermal subsidence	Global eustatic	In-plane stress	Glacioeustatic, tectonic, sedimentary
1st	10^8	|				
2nd	10^7	|		|	|	
3rd	10^6		|	|		
4th	10^5			|	|	|
5th	10^4					|

shallowing-upward units, termed parasequences, resulting from fourth/fifth-order relative sea-level changes (Table 4.1, see later section). The mechanisms behind the relative sea-level changes, noted on Table 4.1, are the subject of much debate, but that is beyond the scope of this chapter.

Sequence stratigraphy has been largely developed from, and applied to, siliciclastic depositional systems but there are some important differences between siliciclastic and carbonate systems in terms of their response to relative sea-level changes (see Sarg, 1988; Schlager, 1991; Hunt & Tucker, in press), and

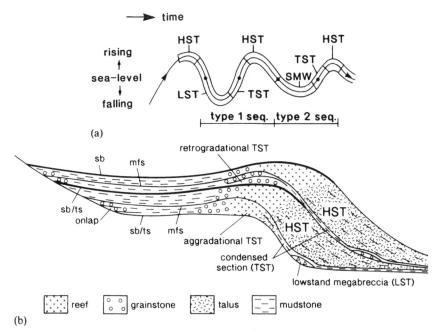

Fig. 4.6 (a) Schematic representation of third-order relative sea-level curve (on the time-scale of 1–10 Ma) producing a type 1 sequence with lowstand (LST), transgressive (TST) and highstand (HST) systems tracts, when the relative sea-level fall is strong and the whole shelf is exposed, and a type 2 sequence with shelf margin wedge (SMW), TST and HST, formed by a minor relative sea-level fall. (b) Schematic sequence stratigraphic model for a carbonate rimmed shelf. Two sequences are shown: (i) an aggradational TST, produced by a reef which was able to keep pace with the relative sea-level rise; (ii) retrogradational shelf-margin grainstone parasequences in the TST. In both sequences the lowstand is represented by just a small apron of megabreccia, derived from collapse of the earlier highstand margin during the sea-level fall. Toe-of-slope megabreccias can also be produced by oversteepening of the shelf margin during the TST and HST. sb, sequence boundary; mfs, maximum flooding surface; ts, transgressive surface.

these do have a consequence for early diagenesis. In siliciclastic systems, type 1 and type 2 sequence boundaries are distinguished. The first is produced by a relative sea-level fall to below the offlap-break (which for clastic shelves is at a depth of around 10–20 m), and then a LST is established. The type 2 sequence boundary is generated by a less severe relative sea-level fall to close to the offlap-break, and then a SMW systems tract is developed. In carbonate systems, the water depth at the margin of a rimmed shelf is mostly <10 m, so that almost any relative sea-level fall, except the most minor, will produce a type 1 sequence boundary. (It is for this reason that SMWs are rarely developed in carbonate shelf sequences.) The importance of this in terms of carbonate diagenesis is that for most relative sea-level falls, the whole shelf or all of the inner ramp will be exposed and thus subject to subaerial, surface-related diagenesis, meteoric or marine. However, this does not mean that in car-

bonate systems every relative sea-level fall produces a sequence boundary. Exposure, usually of a much shorter time interval than that occurring at a sequence boundary, has taken place at many parasequence boundaries. In limestone successions with parasequences, the sequence boundaries are defined on the stacking patterns of the parasequences (e.g. Goldhammer *et al.*, 1990; also see later section and Fig. 4.10).

Compared with siliciclastics, relatively small volumes of carbonate sediment are generally deposited during the LST since the area for carbonate production is considerably reduced. There is also a much greater range of responses of a carbonate system to a relative sea-level rise. This is due to the fact that in many cases carbonate sedimentation rates are able to keep up with and even exceed the rates of relative sea-level rise. Thus, aggradation and even progradation of shelf rims/inner-ramp sand bodies can take place during the TST. In other instances, notably

GREENHOUSE PERIOD

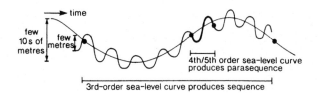

ICEHOUSE PERIOD

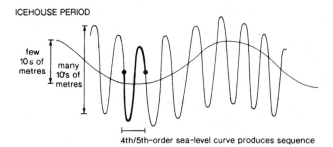

Fig. 4.7 Schematic third-order and fourth/fifth-order relative sea-level curves for greenhouse and icehouse periods. Third-order sequences with fourth/fifth-order parasequences are typical of greenhouse periods, whilst fourth/fifth-order sequences are produced during icehouse periods.

where local environmental factors play a role or when the relative sea-level rise really is too rapid, retrogradation of the shelf margin ramp shoreline facies takes place, which is the typical response of a siliciclastic system. Substantial thicknesses of limestone are commonly deposited during the TST, with the significance that they experience extended periods of marine diagenesis. Also compared with siliciclastic systems, much greater amounts of carbonate sediment are produced during the highstand (the concept of highstand shedding), and shelf-margin progradation is most pronounced at this time.

Carbonate diagenesis and relative sea-level falls (LST)

With a relative sea-level fall, and the establishment of a LST, there is a basinward shift of the groundwater zones within a carbonate platform (see Figs 4.8 and 4.9). *If the climate is arid*, then hypersaline waters are likely to be developed in lagoons, lakes, salinas and sabkhas on the shelf or inner ramp. The waters here may be derived from the sea during major storms, if the relative sea-level fall is not too great, or from continental runoff. The descent of the brines into the subsurface is likely to lead to reflux-type dolomitization of the TST–HST carbonates of the earlier sequence (Fig. 4.5) and possibly cementation by gypsum–anhydrite–halite. There are many ancient examples of massive dolomites associated with evaporites which formed during a rela-

tive sea-level fall (e.g. the Zechstein of western Europe, the Permian of the Delaware Basin and the Silurian of the Michigan Basin; see Kendall, 1989 for a review, and Tucker, 1991b). Karstification of the previously-deposited sequence is likely to be minor in an arid setting.

If the climate is humid, then the exposed highstand sediments will be subject to freshwater diagenesis, with aragonite dissolution, cementation by calcite spar, and sediment stabilization to low-Mg calcite, taking place in vadose and phreatic zones. Soils such as calcretes may form, and laminated crusts and karstic surfaces beneath the soil upon the carbonate rock substrate. In the near-surface, extensive karst systems may develop within the earlier HST and TST carbonates, with rejuvenation of the karst occurring as relative sea-level falls, especially if the fall is episodic. Within the shallow subsurface, the meteoric groundwater zone will migrate seawards during the lowstand, and dolomitization of highstand sediments and carbonates of earlier sequences could take place in the meteoric-marine mixing-zone, or within the zone of circulating marine pore-fluids ahead of the mixing-zone (Fig. 4.5).

If the carbonate platform is of the rimmed-shelf type and a lowstand wedge develops during the LST (not common in carbonate systems), or the platform is a ramp and a sand body develops at the new, more basinward shoreline, then the lowstand sediments deposited will be subjected to marine diagenesis. As noted earlier, this is likely to be cementation

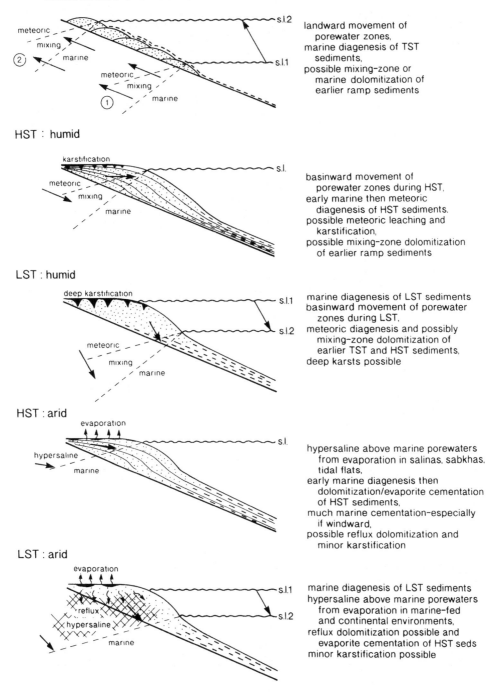

TST : arid, windward–much marine cementation
humid, leeward–little marine cementation

landward movement of
  porewater zones,
marine diagenesis of TST
  sediments,
possible mixing-zone or
  marine dolomitization of
  earlier ramp sediments

HST : humid

basinward movement of
  porewater zones during HST,
early marine then meteoric
  diagenesis of HST sediments,
possible meteoric leaching and
  karstification,
possible mixing-zone dolomitization
  of earlier ramp sediments

LST : humid

marine diagenesis of LST sediments
basinward movement of porewater
  zones during LST,
meteoric diagenesis and possibly
  mixing-zone dolomitization of
  earlier TST and HST sediments,
deep karsts possible

HST : arid

hypersaline above marine porewaters
  from evaporation in salinas, sabkhas,
  tidal flats,
early marine diagenesis then
  dolomitization/evaporite cementation
  of HST sediments,
much marine cementation—especially
  if windward,
possible reflux dolomitization and
  minor karstification

LST : arid

marine diagenesis of LST sediments
hypersaline above marine porewaters
  from evaporation in marine-fed
  and continental environments,
reflux dolomitization possible and
  evaporite cementation of HST seds
minor karstification possible

Fig. 4.8 Models for the diagenesis of a carbonate ramp in relation to its depositional systems tracts under arid and humid climates.

TST – unconfined aquifer

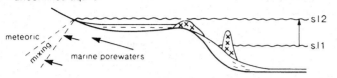

landward movement of porewater
   zones,
marine diagenesis of TST sediments,
possible mixing–zone–related or marine
   dolomitization of earlier shelf sediments

arid windward – much marine cementation
humid leeward – little marine cementation

HST – humid, unconfined aquifer

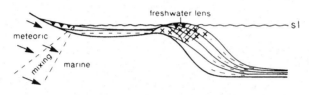

marine diagenesis of HST sediments
basinward movement of porewater
   zones during late HST sea-level fall
meteoric lens developed at shelf margin
possible meteoric leaching, karstification
   and mixing–zone–related dolomitization
   of HST and earlier sediments

LST –humid, unconfined aquifer

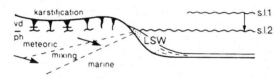

basinward movement of porewater
   zones during LST,
meteoric diagenesis and possibly
   mixing–zone–related dolomitization
   of earlier TST and HST sediments,
deep karsts possible,
marine diagenesis of LSW sediments

HST – arid, unconfined aquifer .

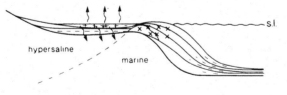

much marine cementation of HST
   sediments –especially if windward
   margin
hypersaline waters reflux from
   evaporating lagoons
early marine diagenesis then
   dolomitization

LST – arid, unconfined aquifer

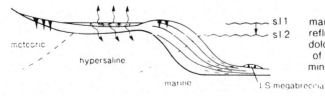

marine diagenesis of LST sediments
reflux of hypersaline waters
dolomitization and evaporite cementation
   of earlier HST and TST sediments
minor karstification

Fig. 4.9 Models for the diagenesis of a carbonate shelf in relation to its depositional systems tracts under a humid climate.

if the climate is arid and the platform orientation is windward. These LST carbonates are unlikely to suffer extensive meteoric diagenesis or karstification, since they will be buried in marine pore-fluids, once relative sea-level begins to rise.

During the lowstand in a carbonate system, it is not uncommon for megabreccias to be shed from a shelf margin into the basin to form allochthonous wedges (equivalent to a lowstand fan of siliciclastic systems). The blocks in such deposits consist of highstand facies and typically show evidence of subaerial exposure (meteoric cements, leached fossils, karstic vugs, etc.), reflecting the sea-level fall before the shelf-margin collapse. Once deposited, marine internal sediments and cements may well be precipitated within the blocks, and their outer surfaces bored, encrusted and mineralized.

Carbonate diagenesis during relative sea-level rises (TST)

During a relative sea-level rise, groundwater zones move landward through a carbonate platform (see Figs 4.8 and 4.9). The sediments deposited during the TST will be subject to marine diagenetic processes, which act principally at the sea-floor and up to about 1 m below. However, since relative sea-level is rising, these TST sediments (along with those of the LST already deposited in this sequence) will be buried with marine pore-fluids, at least until any major flushing by meteoric water during a subsequent lowstand, if the climate is humid. The consequence of this is that further diagenetic changes are likely to be severely curtailed, until such time as the pore-waters change or burial into the zone of compaction takes place. Sediments of the TST could well have abundant marine cements, but if they move into the shallow-burial, stagnant marine diagenetic environment relatively rapidly, then they will not suffer any further cementation (which is dependent on fluid-flow rates) and porosity loss. A small amount of cementation may also be sufficient to delay burial compaction, and this would enhance the rock's reservoir potential, especially if oil generation within the basin were relatively early.

In the last few years, it has been suggested that much early dolomitization may be the result of seawater being pumped through carbonate platforms, in some cases with the seawater chemistry modified a little, as through sulphate reduction or meteoric water dilution. Mechanisms suggested for driving the water through the sediments are oceanic tides and currents and Kohout convection. However, one further way is a relative sea-level rise (Fig. 4.5). In this situation, the rising sea-level causes the marine pore-water zone to push the mixing-zone and meteoric zone ahead of it, landwards, through the TST but especially through the LST–TST–HST sediments of the previous sequence. The active circulation in the marine pore-water zone, and in the vicinity of the mixing-zone, could lead to pervasive dolomitization. Sulphate contents could well have been reduced within the marine pore-waters from earlier microbial sulphate reduction during the lowstand, when the circulation of pore-waters would have been less vigorous and organic matter degradation would have taken place in the more stagnant conditions.

Carbonate diagenesis during relative sea-level highstands (HST)

When relative sea-level reaches a highstand, then the main feature of carbonate sedimentation is aggradation and progradation. Carbonate platforms prograde to their greatest extents during the HST, and pronounced clinoforms may be produced. During the highstand, the type of carbonate diagenesis is again strongly dependent on the climate (Figs 4.8 and 4.9). Marine cementation may well be important along the shelf margin or ramp shoreline, especially if the climate is arid and the platform orientation is windward. However, as a result of the progradation, which is typical of HSTs, the marine sediments soon come under the influence of the supratidal and subaerial diagenetic environments.

In arid regions, sabkhas, salinas and hypersaline lagoons are likely to be present behind tidal flats. Polygonal structures, surficial crusts, tepees and associated pisoids, will be present on high intertidal–supratidal flats as a result of calcite/aragonite precipitation and cementation from the evaporation of seawater. Evaporative dolomitization and dolomite precipitation will occur in the supratidal environment and some very shallow subsurface intertidal–shallow subtidal carbonates may be dolomitized. There may be some reflux dolomitization, but it will not be extensive until there is a relative sea-level fall, allowing deeper penetration of hypersaline brines. Some evaporite cementation of the carbonates may take place. All these processes of arid-region, surface-related, subaerial-supratidal diagenesis can cause much sediment disruption, cementation and alteration.

In humid regions, there will be a basinward movement of groundwater zones during highstand progradation, with the driving force being meteoric water recharge (Figs 4.8 and 4.9). This process will increase in effect during the later stages of the HST when sea-level begins to fall. With a rimmed shelf, a meteoric lens will develop at the shelf margin if the reef/shoal sediments become subaerial (Fig. 4.9). Meteoric diagenesis will affect the sediments and this may well result in karstic surfaces, laminated crusts and paleosols such as calcretes. Deeper karst systems will not develop until the meteoric water-table drops substantially through a relative sea-level fall. Meteoric leaching and cementation of sediments will occur. Dolomitization may take place in the meteoric–marine mixing-zone below the shelf and/or in the circulating marine water ahead of the mixing zone, and below the freshwater lens at the shelf margin.

Thus, during the highstand, the main diagenetic pattern will be an initial marine diagenesis replaced in time either by supratidal diagenesis and evaporative dolomitization or meteoric dissolution, cementation and mixing-zone-related dolomitization, depending on the climate. Figure 4.2a shows an oolite which was cemented in the marine environment (by an acicular calcite cement) and then subjected to meteoric diagenesis, with leaching of the originally aragonitic ooids and bioclasts, partial leaching of the marine cement (suggesting it was high-Mg calcite originally) and some cementation by calcite spar.

Carbonate diagenesis on the parasequence scale

The discussion in preceding sections has been concerned with diagenesis at the level of sequences, but many carbonate sequences consist of smaller-scale units, parasequences, produced by relative sea-level changes on a shorter time-frame (10 000–100 000 years mostly). Some sequences consist of many tens or even hundreds of metre-scale parasequences and in some instances they are bundled into packets of four to six. Although there is a great variation in facies within parasequences from different formations, they nearly all show shallowing-upward trends. Parasequences are generally the result of transgressive–regressive events, with most of the sediment deposited during the stillstand/regressive interval. Most carbonate parasequences have an upper surface recording emergence, but the length

of time represented here is generally relatively short (a few thousand years perhaps), especially compared to many sequence boundaries. The repetition of parasequences may be the result of allocyclic processes, such as orbital forcing in the Milankovitch band (e.g. Goldhammer *et al.*, 1990) or jerky subsidence, or autocyclic processes, such as the Ginsburg tidal-flat progradation/loss of carbonate factory model (see Tucker & Wright, 1990 for discussion).

The various points developed above in the discussions of diagenesis during TST and HST can be applied to parasequences. Thus the transgressive parts of a parasequence would be expected to show marine diagenetic textures, whereas the regressive part would show the effects of surface-related diagenesis, such as palaeokarstic surfaces, laminated crusts, meniscus cements, leached aragonitic bioclasts and calcretes, if the climate was humid and meteoric diagenesis was operative, or supratidal/evaporative dolomite, tepees, crusts, pisoids and botryoids, and displacive gypsum, if the climate was arid.

Parasequences commonly display systematic vertical changes in thickness and facies through a sequence, and in association with these, there may be systematic changes in the type and degree of diagenesis. The stacking patterns of the parasequences reflect the longer-term third-order relative sea-level curve, and so vary with the systems tracts of the sequence. Parasequences developed during TST, when the third-order relative sea-level curve is rising, typically show a thickening-upward trend, and the proportion of subtidal facies to intertidal facies increases (Fig. 4.10). These two features reflect the increased accommodation space for each successive parasequence. Parasequences developed when the third-order relative sea-level curve is falling, that is, during late HST and early LST, will show a thinning-upward trend, and intertidal facies will dominate over subtidal facies (Fig. 4.10). These two features reflect the decreasing amount of accommodation space for each successive parasequence. Parasequences deposited during the highstand should be of similar thickness, as should those of the lowstand, but those of the HST should generally be thicker than those of the LST. It is possible, of course, for no parasequences to be deposited during the lowstand, if sea-level falls below the platform margin for several or many of the fourth/fifth-order sea-level fluctuations (the concept of 'missed beats'). This would generate a prominent sequence boundary.

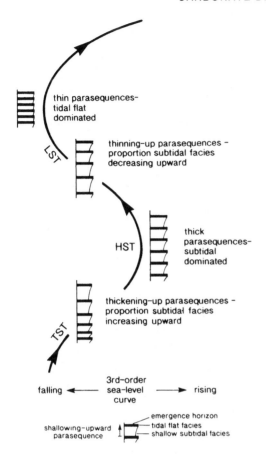

thin parasequences-
tidal flat
dominated

LST

thinning-up parasequences -
proportion subtidal facies
decreasing upward

HST

thick
parasequences-
subtidal
dominated

thickening-up parasequences -
proportion subtidal facies
increasing upward

TST

falling ◄———— 3rd-order
sea-level
curve ————► rising

shallowing-upward
parasequence

emergence horizon
tidal flat facies
shallow subtidal facies

Fig. 4.10 Schematic stacking patterns for parasequences deposited on a carbonate platform subjected to high-frequency (fourth/fifth-order, 10 000–100 000 year), low amplitude sea-level changes against a background of lower frequency (third-order, 1–10 m.y.), higher amplitude sea-level change. Parasequences themselves do vary a great deal in their facies development, but within one sequence or one systems tract, they usually have a similar motif. Individual parasequences generally show a lateral facies variation across a platform, with the proportion of tidal flat facies increasing relative to the subtidal facies towards the platform interior.

In terms of diagenesis, parasequences of the TST and third-order sea-level rise should show much more evidence of marine diagenesis than those of the third-order sea-level fall (Fig. 4.11). Parasequences of the third-order sea-level fall should show the effects of surface-related diagenesis to a greater extent than those of the TST. Thus, for example, peritidal dolomitization should be more extensive in the parasequences of the third-order sea-level

fall, compared to those of the third-order sea-level rise. Total dolomitization of the late HST/LST parasequence set could take place in this way. Parasequences of the lowstand may be intensely disrupted and altered by supratidal/subaerial processes: the tepeeization of LST parasequences in the Triassic Latemar platform of the Dolomites would be a good example of this (Goldhammer *et al.*, 1990).

In some instances, the carbonate platforms where parasequences were deposited were very extensive (the Cambrian/Ordovician of North America and China for example). With these parasequences, in the same way that there are vertical variations, there are lateral variations in the degree of surface-related diagenesis, with the more landward parts of the platform showing more meteoric effects, if the climate were humid, or more supratidal dolomitization, if the climate were arid (Fig. 4.11). On a broad scale, the 'boundary' between intense, surface-related diagenesis and less affected carbonates should onlap in the TST and offlap for the HST/LST parasequences (Fig. 4.11).

Carbonate diagenesis and sequence stacking patterns

The diagenetic models presented in Figs 4.8 and 4.9 are based on a single sequence, although as has been noted on several occasions in the preceding discussions, the diagenesis taking place during the deposition of one sequence may affect the sequence below. This situation is particularly significant with regard to meteoric diagenesis and dolomitization, when diagenetic alteration of highstand sediments will be most intense during deposition of the lowstand sediments of the next sequence. However, this approach of considering diagenesis within the framework of a sequence should be taken a stage further since the longer term path of diagenesis for a sequence does depend on the sequence stacking pattern (i.e. the type of megasequence).

Three end-member sequence stacking patterns can be recognized (Fig. 4.12), with the controlling factor being the relationship between the third-order (1–10 m.y.) relative sea-level change that produced the sequence and the second-order (10–100 m.y.) relative sea-level change that produced the sequence set. Where the third-order relative sea-level change takes place against a background of little or no second-order relative sea-level change, then a progradational sequence set is produced:

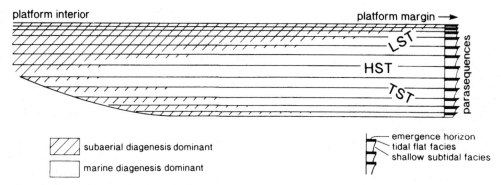

Fig. 4.11 Generalized scheme indicating the lateral and vertical variations in carbonate diagenesis for parasequences of different systems tracts. The type of subaerial (or surface-related) diagenesis depends very much on the climate, with palaeokarstic surfaces and meteoric cements characterizing a humid climate, and supratidal dolomitization and tepeeization typical of an arid climate. This subaerial diagenesis only affects the upper parts of platform interior parasequences deposited during the third-order sea-level rise (TST), but it may affect the whole of the parasequences deposited during the third-order sea-level fall (late HST/LST). Parasequences of the lowstand may be severely affected by near-surface diagenetic processes, as in mega-tepee horizons. The sequence boundary would lie within these LST parasequences, but it need not be one particular horizon.

where the second-order relative sea-level change is a moderate rise, then an aggradational sequence set is produced, and where the second-order relative sea-level change is a strong rise, a retrogradational sequence set is produced. Examples of these different stacking patterns are: for a progradational sequence set, the Upper Jurassic/Lower Cretaceous carbonate platforms of the Baltimore Canyon (subsurface eastern USA continental shelf), of the Neuquen Basin (Argentina) and of the Subalpine Chains (French Alps); for an aggradational sequence set, the Lower Carboniferous ramp carbonates of South Wales and the Middle Jurassic oolites of the Wessex Basin; and for a retrogradational sequence set, the Upper Cretaceous platforms of the Spanish Pyrenees.

Progradational sequence sets and diagenesis

Where sequences are arranged into a prograding sequence set, then near-surface, shallow-burial, post-marine diagenesis operates over a very long period of time. The sequences, by offlapping, are not buried to any great extent, until there is a change in the second-order relative sea-level curve or some major tectonic event. If the climate is humid, then there is the potential for extensive meteoric leaching and deep karsts. Mixing-zone-related dolomitization may also occur. If the climate is arid, then pervasive evaporative and reflux

dolomitization is likely to occur. The early diagenesis of all the sequences in the set will be similar, but with more advanced stages reached by the sequences closer to the basin margin. The sequence set as a whole will be capped by a zone of sequence boundaries, to which the early diagenetic events will be related. With the prolonged period of near-surface, shallow-burial diagenesis, most of the sediments are going to be well-lithified before burial to greater depths. As a result of this, mechanical compaction is unlikely to take place and chemical compaction will be delayed.

Aggradational sequence sets and diagenesis

Where sequences are arranged into an aggradational sequence set, then the early diagenetic processes (marine and/or meteoric) will be followed by burial diagenetic processes, with the former determining the path of the latter. Mechanical and chemical compaction is likely to be a major burial process since, in many of the limestone units, only partial near-surface lithification will have taken place. In this vertical stack of carbonate platforms, there will be a strong possibility of confined aquifers, such as porous HST grainstones separated by TST mudstone aquicludes. These may well permit the ingress of meteoric waters into the sedimentary package from hinterland recharge areas. Karst porosity developed at sequence boundaries may be occluded by

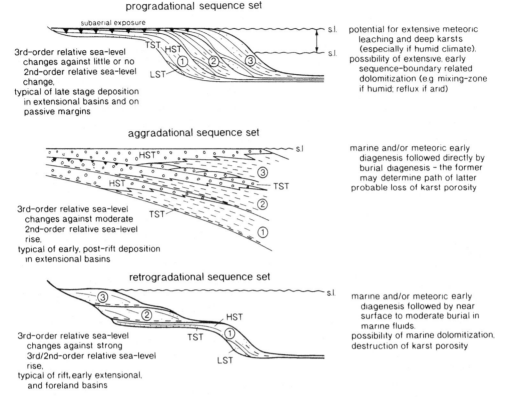

Fig. 4.12 Models for the diagenesis of carbonate platforms in relation to different sequence stacking patterns. The stacking patterns are a reflection of the relationship between the third-order (1–10 m.y.) relative sea-level changes which produce the sequence, and the second-order (10–100 m.y.) relative sea-level changes which produce the sequence set (megasequence). A progradational sequence set is produced when the third-order relative sea-level change takes place against a background of little or no second-order change. An aggradational sequence set is produced when the second-order change is a moderate rise, and a retrogradational sequence set is produced when the second-order change is a strong rise. The progradational and retrogradational sequence stacking patterns are drawn with carbonate shelves, whereas the aggradational set is drawn with carbonate ramps.

sediments from the succeeding sequence, although rapid burial may preserve some of it.

Retrogradational sequence sets and diagenesis

Where the sequences are arranged into a retrogradational sequence set, then early marine and/or meteoric diagenesis will be followed by shallow to moderate burial in marine pore-waters. As noted earlier, such a situation is likely to lead to a retardation of diagenetic processes until a change of pore-water chemistry or burial collapse of pore networks through compaction takes place. This situation is the optimum for the preservation of primary or reduced primary porosity, especially

where there has been an early partial lithification so that burial compaction is retarded. The backstepping arrangement of the platforms is also likely to result in condensed sections, which are commonly composed of organic-rich mudrocks, being deposited over the earlier sequences. These may act as seals to reservoir units and they may even be source rocks. Also as a function of a retrogradational stacking pattern of platforms, karst porosity developed at a sequence boundary may be destroyed through erosion of the karstic terrain and filling of the caverns during the TST of the next sequence.

Burial overburden of earlier platforms will not be excessive, unless there is an axial source of mud into the basin. Thus burial compaction and associated porosity loss may occur relatively late in the platform's history. The retrogradational arrangement of sequences may lead to extensive dolomitization by marine pore-waters, if this is an important mechanism of dolomitization and there is an efficient system of seawater circulation within the sediments.

In terms of reservoir, seal, source rock and primary porosity retention, carbonate platforms formed during second-order relative sea-level rises have the highest hydrocarbon potential. This is borne out by the important occurrence of major reservoirs in mid-Upper Silurian, mid-Upper Devonian and Upper Cretaceous carbonate platforms, all periods of second-order relative sea-level rise on the Haq *et al.* (1987) global sea-level curve.

Thus, from the preceding discussion and concepts of sequence stacking developed there, it would seem that the patterns of carbonate diagenesis on a large scale should tie in with second-order relative sea-level changes, that is on the scale of tens of millions of years. Carbonate sequence stacking patterns do reflect the type of sedimentary basin and the evolutionary stage of that basin, as indicated in Fig. 4.12.

Carbonate diagenesis and role of associated siliciclastics

The preceding discussion has centered on the stacking of carbonate platforms but in many situations carbonate sequences are interstratified with wholly siliciclastic sequences, or there may have been major influxes of siliciclastics during development of a carbonate platform. The latter is most likely to happen during a lowstand, especially in association with the formation of a type 1 sequence boundary on a rimmed shelf when coarse clastics may be introduced and bypass the platform to constitute lowstand fans. During deposition of carbonates generally, finer-grained siliciclastics are commonly deposited in the basin as hemipelagic muds, and these may form thick units, especially if there is an axial supply.

Siliciclastic sediments associated with or occurring within carbonate sequences will have an important effect on the diagenesis of the carbonate sediments. These effects will be most pronounced during shallow and deep burial. Basinal mudrocks will de-water and this will induce fluid circulation patterns within adjacent carbonates, which may lead to cementation and dissolution. The fluids themselves from mudrocks may be capable of dolomitizing, silicifying and mineralizing nearby carbonates, or more distant shallow-water carbonates if there are permeable slope facies or basin-margin faults to act as conduits. Maturation of organic matter in basinal mudrocks will lead to the generation of P_{CO_2}-rich fluids which may be able to leach calcite and dolomite and create porosity. Sandstones interbedded with carbonates in more proximal locations are likely to provide pathways for the influx of meteoric waters into the carbonate platform. Dissolution of the carbonates may well result, and the sandstones themselves may be subject to carbonate cementation.

The role of siliciclastics in carbonate diagenesis is a topic for further discussion elsewhere, as is the importance of structural elements, notably faults and fractures, on fluid migration and resulting diagenesis. Nevertheless, it should be remembered that mudrocks are the dominant lithology filling sedimentary basins, and their composition, chemistry and burial history must be taken into account, particularly when considering the shallow and deep-burial diagenesis of carbonate rocks.

Carbonate diagenesis through the Phanerozoic: broad trends and the first-order sea-level curve

Two broad aspects in which carbonate diagenesis varies through time have already been mentioned, namely the variations in mineralogy and fabric of marine cements, and the variations in diagenetic potential of carbonate sediments, reflecting their original mineralogy.

The composition of marine cements has varied through the Phanerozoic (Fig. 4.13), with periods when aragonite and high-Mg calcite were dominant (Cambrian, Pennsylvanian–Triassic, Cenozoic) and periods when calcite with low Mg content was the main cement (mid-Palaeozoic, Jurassic/Cretaceous). The pattern of marine cements ties in with the first-order sea-level curve (i.e. on the scale of hundreds of millions of years, Table 4.1), with aragonite and high-Mg calcite being the precipitates at times of global sea-level lowstand, and calcite at times of global sea-level highstand (Fig. 4.13; Sandberg, 1985; Tucker, 1992). Changes in seawater chemistry, principally in P_{CO_2} and Mg : Ca ratio, brought about by changes in rates of sea-floor spreading and subduction, are considered the underlying controls.

The mineralogy of Phanerozoic ooids, which determines the diagenetic potential of oolitic sediments, varies in the same manner as the marine cements, correlating with the first-order sea-level curve and reflecting seawater chemistry (Fig. 4.13; Sandberg, 1983; Wilkinson *et al.*, 1985). However, there are some exceptions to the trend, notably in the Upper Jurassic where aragonitic ooids were relatively common (e.g. the Smackover of the Gulf Coast subsurface, Fig. 4.3a; Swirydczuk, 1988). These probably reflect local conditions of elevated Mg : Ca ratio. Oolites make excellent hydrocarbon reservoirs but the type of porosity does depend on original ooid mineralogy, as well as subsequent diagenesis. Aragonitic ooids, with a high-diagenetic potential for dissolution, give rise to oomolds (Figs 4.2a and 4.3a). Some fracturing may be necessary to increase the effective porosity. Calcitic ooids are much more stable (low diagenetic potential) and

where they form reservoir rocks the porosity tends to be reduced primary intergranular, as a result of a little early cementation, useful to inhibit burial compaction (as in Fig. 4.3b). Porosity may be generated in oolites by dolomitization and associated calcite dissolution, or karstification.

There is apparently a secular variation in the distribution of dolomites through time (Fig. 4.13; Given & Wilkinson, 1987), with abundances coinciding with the highstands of the first-order sea-level curve (also the periods of calcite precipitation). If this trend is correct, it again suggests an over-riding geotectonic control on carbonate diagenesis; in this case it may reflect the more efficient circulation of seawater within carbonate platforms at times of relative sea-level highstand.

Carbonate diagenesis and sequences through the Phanerozoic

The nature of depositional sequences themselves does vary through the Phanerozoic. In the concepts of sequence stratigraphy, as expounded by Van Wagoner *et al.* (1988, 1990), sequences are defined on the basis of unconformities and these are produced by relative sea-level falls. Sequence stratigraphy has evolved from seismic stratigraphy and most of the sequences described in the papers of the last few years, in the subsurface from seismic, core and logs, and at outcrop, are the product of third-order relative sea-level changes, that is, on the time-scale of 1–10 m.y. (10^6–10^7 years, Table 4.1) and with an amplitude of many tens of metres (see Figs 4.6 and 4.7). These sequences are typically 50–200 m thick, are divisible into the various systems tracts discussed earlier (Fig. 4.6), and consist of metre-scale shallowing-upward parasequences. The latter are the product of smaller-scale (less than 10 m amplitude) fourth/fifth-order relative sea-level changes on the time-scale of 10^5–10^4 years (Table 4.1, Figs 4.7 and 4.10).

The amplitude of the various orders of relative sea-level change has varied through time, with one of the main factors controlling the higher orders of sea-level change (fourth/fifth-order), being the presence or absence of polar ice caps. The Phanerozoic has been divided into various icehouse and greenhouse periods on the basis of the Earth's glacial record (see Fig. 4.13; Veevers, 1990). During times of no polar ice caps or only mountain glaciation, that is, greenhouse times, the fourth/fifth-order relative sea-level changes are on the scale of a few

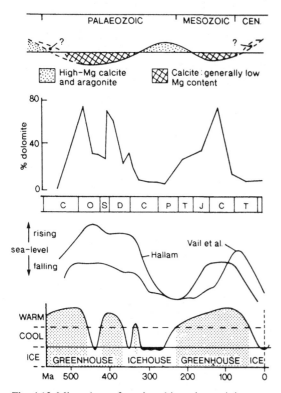

Fig. 4.13 Mineralogy of marine abiogenic precipitates (ooids and cements) and abundance of dolomites through the Phanerozoic compared with the first-order global sea-level curve and the icehouse–greenhouse periods (data from Sandberg, 1983; Given & Wilkinson, 1987; Veevers, 1990).

metres whereas the third-order relative sea-level changes are on the scale of many tens of metres. These are the conditions for the generation of the now well-documented third-order sequences containing fourth/fifth-order parasequences (Fig. 4.10). However, during icehouse times, and the Earth is in one now, the amplitude of the fourth/fifth-order sea-level changes is much greater than any third-order relative sea-level change (Fig. 4.7). Sea-level changes in excess of 100 m occurred during the last ice age. These sea-level changes are also very rapid. It is generally accepted that these sea-level changes are the result of the waxing and waning of the polar ice caps in response to changes in insolation brought about by orbital-forcing (Milankovitch rhythms). These glacioeustatic fourth/fifth-order sea-level changes have a greater amplitude than the longer-term third-order relative sea-level changes and themselves generate unconformities upon rimmed shelves and ramps as a result of the drastic falls in sea-level. Therefore, during icehouse times, sequences are produced by the glacioeustatic sea-level changes, and these *fourth/fifth-order sequences* are much thinner than the third-order sequences, which are typical of the greenhouse times. Thus, although the Quaternary carbonate sequences of Barbados and the Bahamas are thin (5–30 m) and were deposited in a short time (the order of 10 000–100 000 years) when compared to the sequences of the Mesozoic say, they are all true sequences in the accepted concepts of sequence stratigraphy, since they are bounded above and below by unconformities. In terms of the time involved for deposition, the fourth/fifth-order sequences are equivalent to the parasequences which make up third-order sequences. In terms of thickness, the fourth/fifth-order sequences are generally thicker than the parasequences, at least in part because of the greater amount of accommodation space created by the higher amplitude sea-level change. Apart from the Quaternary, fourth-order sequences will also be present in the Pennsylvanian/Permian (the time of major glaciation in Gondwanaland), in the late Ordovician/early Silurian (glaciation in northern Africa particularly) and in the latest Precambrian (glaciations on all continental land masses).

This distinction between third-order and fourth/fifth-order sequences has important consequences for carbonate deposition and diagenesis. In terms of carbonate shelves, rimmed shelves will be better developed during icehouse times, when the substan-

tial, short-term sea-level fluctuations promote carbonate deposition at shelf margins and accentuate the rim-lagoon topography. During greenhouse times, the metre-scale, short-term sea-level changes against a background of larger-scale third-order change, will favour the development of aggraded shelves with little topographic difference across the shelf; metre-scale shallowing-upward cycles (parasequences) will dominate the succession (Fig. 4.14). Very extensive epeiric platforms are also likely to be better developed during greenhouse periods.

There will be differences in the pattern of carbonate diagenesis between the fourth/fifth-order and third-order sequences. Under a humid climate, the high amplitude sea-level changes producing the fourth/fifth-order sequences lead to the formation of deep karsts (Fig. 4.14). They develop from the sequence boundary during the lowstand. The deep karst of the Bahama platform (the blue holes) is a manifestation of the dramatic sea-level changes of the last million years. Deep karst also occurs in Permian/Carboniferous strata of the Delaware Basin, as in the San Andres Formation (Craig, 1990), a time of fourth/fifth-order sequences. With third-order sequences, minor karst is commonly developed at the parasequence boundaries (Fig. 4.14), and more prominent karstic surfaces may form at the sequence boundaries.

The rapid and high-amplitude relative sea-level changes of icehouse times will lead to more rapid pore-fluid movements through carbonate platforms than during greenhouse times. During an icehouse period, meteoric fluids will be more penetrating. The flushing of meteoric water into a platform during a lowstand will lead to deep leaching and grain dissolution. By way of contrast, in greenhouse times, much of the diagenesis will take place near-surface, since sea-level changes are only on the scale of a few metres, until a major relative sea-level fall terminates the sequence and permits deeper ground water movements. During icehouse periods, it is probable that some pore-fluid migrations will be too fast to allow diagenetic reactions to take place. During the TST, the rapid sea-level rise will probably preclude any extensive marine cementation or seawater dolomitization. There may well be insufficient time for significant mixing-zone-related dolomitization to take place during a HST or LST. The rapid sea-level rise from 20 000 to 5000 years ago and the establishment of mixing-zones only recently could account for the general paucity of mixing-zone-related dolomites forming at the present time.

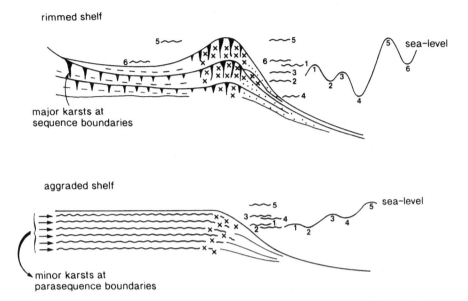

Fig. 4.14 Schematic models for a carbonate rimmed shelf, typical of (but not unique to) icehouse periods when fourth/fifth-order relative sea-level changes have a high amplitude (many tens of metres), and an aggraded shelf, typical of (but not unique to) greenhouse periods when fourth/fifth-order relative sea-level changes are subdued (only a few metres). Under a humid climate (as depicted here), meteoric diagenesis has a much greater effect on rimmed shelves than on aggraded shelves.

Summary

In this review, the patterns of carbonate diagenesis have been related to relative sea-level changes and climate, and various models have been put forward placing the diagenesis into the context of sequence stratigraphy and systems tracts. It is suggested that the diagenetic processes acting in the various systems tracts are different and that they can be predicted. The popular mechanisms of dolomitization can also be tied into sequence stratigraphy. Sediments are liable to be affected by diagenetic processes throughout their existence so that, in addition, diagenesis has to be considered on the scale of sequence sets. On a large scale, patterns of carbonate diagenesis should relate to sequence-stacking patterns and second-order relative sea-level changes.

There are also major changes in the pattern of carbonate diagenesis through time. It is now well-established that the mineralogy of marine cements has varied through the Phanerozoic, reflecting fluctuations in seawater chemistry. The diagenetic potential of carbonate sediments, mainly dependent on aragonite content, has varied through time as the composition of abiogenic and biogenic grains has changed in response to seawater chemistry and organism development. The magnitude of fourth/fifth-order relative sea-level changes has varied through time in response to icehouse–greenhouse periods and this has resulted in variations in the types of platform developed and in the degree of meteoric diagenesis and karstification.

Acknowledgements

I am most grateful to Jim Marshall for comments on the manuscript, Karen Atkinson for drafting the figures, Gerry Dresser and Alan Carr for photography and Le Chinaillon (4/91) for inspiration.

References

Craig, D.H. (1990) Yates and other Guadalupian (Kazanian) oil fields, US Permian Basin. In: *Classic Petroleum Provinces* (Ed. J. Brooks). Geol. Soc. Spec. Publ. 50, 249–263.

Given, R.K. & Wilkinson, B.H. (1987) Dolomite abundance and stratigraphic age: constraints on rates and mechanisms of Phanerozoic dolostone formation. *J. Sedim. Petrol.* **57**, 1068–1078.

Goldhammer, R.K., Dunn, P.A. & Hardie, L.A. (1990) Depositional cycles, composite sea-level changes, cycle stacking patterns, and the hierarchy of stratigraphic forcing: examples from Alpine Triassic platform carbonates. *Geol. Soc. Am. Bull.* **102**, 535–562.

Haq, B.U., Hardenbol, J. & Vail, P.R. (1987) Chronology of fluctuating sea-levels since the Triassic. *Science* **235**, 1156–1167.

Hunt, D. & Tucker, M.E. (in press) The sequence stratigraphy of carbonate shelves with an example from the mid-Cretaceous of SE France. In: *Sequence Stratigraphy and Facies Associations* (Eds H.W. Posamentier, C.P. Summerhayes, B.U. Haq & G.P. Allen). Spec. Publ. Int. Ass. Sediment. 18.

Kendall, A.C. (1989) Brine mixing in the Middle Devonian of western Canada and its possible significance to regional dolomitization. *Sedim. Geol.* **64**, 271–285.

McIlreath, I.A. & Morrow, D.W. (Eds) (1990) *Diagenesis.* Geosci. Can. Rep. Ser. 4.

Moore, C.H. (1989) *Carbonate Diagenesis and Porosity.* Elsevier: Amsterdam.

Palmer, T.J., Hudson, J.D. & Wilson, M.A. (1988) Palaeoecological evidence for early aragonite dissolution in ancient calcite seas. *Nature* **335**, 809–810.

Purser, B.H. (1978) Early diagenesis and the preservation of porosity in Jurassic limestones. *J. Petrol. Geol.* **1**, 83–94.

Sandberg, P.A. (1983) An oscillating trend in Phanerozoic non-skeletal carbonate mineralogy. *Nature* **305**, 19–22.

Sandberg, P.A. (1985) Aragonite cements and their occurrence in ancient limestones. In: *Carbonate Cements* (Eds N. Schneidermann & P.M. Harris). Soc. Econ. Paleont. Miner. Spec. Publ. 36, 33–57.

Sarg, J.F. (1988) Carbonate sequence stratigraphy. In: *Sea-level Changes — an Integrated Approach* (Eds C.K. Wilgus, B.S. Hastings, H. Posamentier *et al.*). Spec. Publ. Soc. Econ. Paleont. Miner. 42, 155–181.

Schlager, W. (1991) Depositional bias and environmental change–important factors in sequence stratigraphy. *Sedim. Geol.* **70**, 109–130.

Swirydczuk, K. (1988) The original mineralogy of ooids in the Smackover Formation, Texas. *J. Sedim. Petrol.* **58**, 339–347.

Tucker, M.E. (1991a) *Sedimentary Petrology: an Introduction to the Origin of Sedimentary Rocks.* Blackwell Scientific Publications: Oxford.

Tucker, M.E. (1991b) Sequence stratigraphy of carbonate-evaporite basins: models and application to the Upper Permian (Zechstein) of Northeast England and adjoining North Sea. *J. Geol. Soc. Lond.* **148**, 1019–1036.

Tucker, M.E. (1992) Limestones through time. In: *Understanding the Earth* (Eds R.C.L. Wilson *et al.*) pp. 347–363. Cambridge University Press: Cambridge.

Tucker, M.E. & Bathurst, R.G.C. (Eds) (1990) *Carbonate Diagenesis.* Int. Ass. Sediment. Rep. Ser. 1.

Tucker, M.E. & Wright, V.P. (1990) *Carbonate Sedimentology.* Blackwell Scientific Publications: Oxford.

Van der Baan, D. (1990) Zechstein reservoirs in the Netherlands. In: *Classic Petroleum Provinces* (Ed. J. Brooks). Geol. Soc. Spec. Publ. 50, 379–398.

Van Wagoner, J.C., Mitchum, R.M., Campion, K.M. & Rahmanian, V.D. (1990) *Siliciclastic Sequence Stratigraphy in Well Logs, Cores and Outcrops.* Am. Ass. Petrol. Geol. Meth. Explor. Ser. 7, 45pp.

Van Wagoner, J.C., Posamentier, H.W., Mitchum, R.M. Jr., *et al.* (1988) An overview of the fundamentals of sequence stratigraphy and key definitions. In: *Sea-level Changes — an Integrated Approach* (Eds C.K. Wilgus, B.S. Hastings, H. Posamentier *et al.*). Soc. Econ. Paleont. Miner. Spec. Publ. 42, 39–45.

Veevers, J.J. (1990) Tectonic–climatic supercycle in the billion-year plate-tectonic eon: Permian Pangean icehouse alternates with Cretaceous dispersed-continents greenhouse. *Sedim. Geol.* **68**, 1–16.

Wilkinson, B.H., Owen, R.B. & Carroll, A.R. (1985) Submarine hydrothermal weathering, global eustasy and carbonate polymorphism in Phanerozoic marine oolites. *J. Sedim. Petrol.* **55**, 171–183.

5 Rudist formations of the Cretaceous: a palaeoecological, sedimentological and stratigraphical review

DONALD J. ROSS AND PETER W. SKELTON

Introduction

Reviews of fossil reefs routinely cast some bizarrely shaped bivalves, called 'rudists', in the role of 'reef builders' for most of the Cretaceous (e.g. James, 1984; Fagerstrom, 1987 and references therein). The rudists are portrayed as having dominated 'reefs' for over 40 m.y., between episodes when scleractinian corals were dominant, accompanied by stromatoporoids and calcareous algae (during the Triassic to early Cretaceous) or mainly by calcareous algae (in the Cainozoic)

Some authors suggest also that the rudists achieved their dominance by direct competitive displacement of the corals from reef communities. This hypothesis has been most explicitly advocated by Kauffman & Johnson (1988), who attributed the success of the rudists to their having evolved 'new morphological and ecological features convergent on the bauplan of successful Phanerozoic reef-building taxa'.

The competitive displacement model has been criticized by Scott *et al.* (1990), who demonstrated that mixed rudist–coral associations continued to thrive on some late Cretaceous shelf and platform margins. They argued, instead, that the coral reefs of the early Cretaceous, which flourished mostly on the seaward flanks of platform margins, fell victims to mid-Cretaceous fluctuations in marine productivity coupled with rises in sea-level. By contrast, the rudists, which at first thrived in inner-shelf and platform areas, were held to have been less affected by these environmental perturbations. Thus, it was argued, they came to dominate margins by default, although sometimes accompanied by the remaining coral stocks. In other words, the competitive displacement model was rejected in favour of one involving an independent decline in coral reefs.

It is questionable, in any case, whether the habitats dominated by rudists were environmentally similar to those in which coral reefs thrived at other times. If they were fundamentally different in

character, the demise of coral reefs and the rise of rudist accumulations could simply have reflected changes in conditions and facies patterns on Cretaceous carbonate platforms. In order to avoid interpretative ambiguity here, we prefer not to use the term 'reef'. Following Masse & Philip (1981), the general term 'rudist formations' may be used to include all facies complexes which incorporate build-ups and biostromes dominated by autochthonous or para-autochthonous rudists. To understand these properly, we feel that it is more profitable to synthesize models from analysis of the rudist formations themselves than to attempt to impose extraneous models based upon other organisms from other times and places.

Our purpose here, then, is to survey the character of a large number of rudist formations, based upon our own observations and those of other authors, so to present some general facies models. We start by exploring the autecological attributes of rudists, and some of their associated biota. Then we survey the fabrics of the associated facies, before going on to review patterns of facies geometry and palaeogeographical and stratigraphical variability of rudist formations. Finally, we comment on the implications of our conclusions for hydrocarbon prospectivity.

The biota of rudist formations

Rudists

The rudist bivalves (superfamily Hippuritacea) were sessile epifaunal suspension feeders, which flourished on carbonate-dominated substrata, in shallow marine settings at low latitudes, from the late Jurassic (mid-Oxfordian) to the late Cretaceous (Maastrichtian). Their evolutionary history and systematics were summarized by Skelton (1978).

Unlike colonial corals, rudists grew as discrete individuals. Each had to install itself independently in its habitat. Lacking a foot (Skelton, 1978), and thus any evident means of active mobility following larval settlement, the rudist individual relied upon shell growth to establish itself. There was therefore a close correlation between shell growth form and the nature of the substrate. The different modes of this relationship have been explored by Skelton & Gili (1991) who noted that rudist shells may be classified into three broad palaeoecological morphotypes (Fig. 5.1).

Elevators

The attached valve (AV) margin was wholly involved in upward, substrate-escaping growth. The mean angle of elevation (E) of the maximum (α) and minimum (β) slopes from the horizontal of the attitude of growth of the AV rim (perpendicular to the growth lines) thus tended towards 90° in life (see Fig. 5.1). Elevation might be lessened because of: strongly unidirectional food-bearing currents (Höfling, 1985); non-vertical orientation of the juvenile shell; secondary toppling (with geniculate recovery-growth; Fig. 5.2a—4); or obstruction to upward growth. However, a definitive usual lower limit is set at $E = 45°$. Stabilization was derived from passive implantation of the AV through ambient sedimentation, sometimes assisted by lateral attachment to neighbours (Fig. 5.2a—5, 6). So the morphotype was typically associated with areas of net sediment accumulation, commonly floored by muddy (wackestone to packstone) substrates, perhaps only sporadically swept by storm traction currents. The cylindrico-conical shell forms usually

attributable to this morphotype allowed clustering, with mutual contact, as well as a variety of solitary modes of growth (Fig. 5.2a).

Clingers

A part (or all) of the AV growth margin was continuously deployed in direct overgrowth of the substrate, forming a broad basal area of frictional contact or even attachment. With that part of the shell wall usually more or less horizontally oriented, E was generally less than 45° in life position (Fig. 5.1). The basal area of contact (A) provided stabilization on hard or relatively firm substrates. Maximization of this area of contact produced a generally convex outline (in plan view) to the base of the shell. Thus A approximated to the entire area of the convex polygon (i.e. without re-entrants) mapped onto the substratum from the plan of the shell ('virtual area of support' (A')), and so $A/A' \simeq 1$ (Fig. 5.1). The requirement for a relatively stable substrate meant that mobile sediments, prone to deflation by traction currents, or to sustained accumulation, were generally unsuitable for clingers. Spasmodic sediment accumulation (e.g. influxes of storm-driven sands), however, could be tolerated by many clingers through upward-stepping growth of the basal surface. This effect is revealed in many examples by the interdigitation of foliaceous outgrowths from the AV, and the surrounding sediment. Preferred substrates were normally stable sediments, or, alternatively, hard surfaces (e.g. other shells, hardgrounds) in a variety of settings.

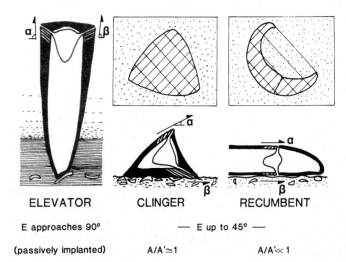

ELEVATOR   CLINGER   RECUMBENT

E approaches 90°   — E up to 45° —

(passively implanted)   A/A'≃1   A/A'≪1

Fig. 5.1 Idealized palaeoecological morphotypes in rudist bivalves (from Skelton & Gili, 1991). Shells are shown in vertical section (black and white) and, for clinger and recumbent, in plan view (above). Maximum and minimum slope of the outer growing margin of the attached valve indicated by α and β, respectively. Elevation index (E) = (α + β)/2. Area of contact with the substrate (A), in the plan views, is shown by close diagonal ornament; virtual area of support (A') is shown by spaced diagonal ornament.

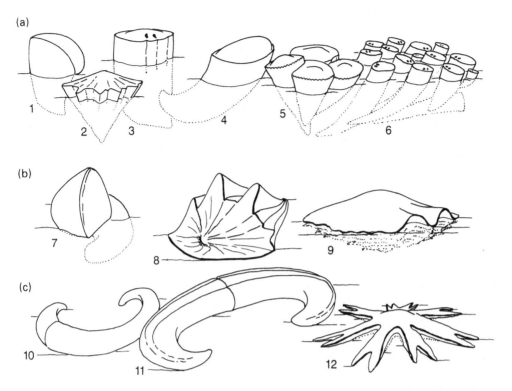

Fig. 5.2 Examples of the three morphotypes. (a) Elevators may be solitary (1–4) or clustered (5, 6) and include many caprotinids and plagioptychids (1), radiolitids (2, 5), hippuritids (3, 6) and multigeniculate forms such as *Dictyoptychus* (4). (b) Clingers overgrew the substrate either laterally (7, 8), or radially (9) and include requieniids (7) and radiolitids (8, 9). (c) Recumbents have arcuate (10, 11) or stellate (12) form, and include caprinids, antillocaprinids and *Sabinia* (10, 11) and radiolitids (12). Parts of shell immersed in sediment are indicated by dotted lines. See Fig. 5.10 for relationships of the families.

Rudists attributable to this morphotype show a wide variety of growth geometries, from spiral (Fig. 5.2b–7) to prone or expanded conical (Fig. 5.2b–8, 9), reflecting the characteristic patterns of growth of the taxonomic groupings to which they belong.

Recumbents

An alternative means of basal stabilization again involved part of the lower valve margin permanently lying flush against the substrate, so that $E < 45°$ (Fig. 5.1). These were not attached (as adults) and, in contrast to the clingers, the area of contact with the sediment (A) was itself of little importance. Rather, growth of a widely projected base, of crescentic or stellate form, produced a relatively large virtual area of support (A'), with broad effective diameters (Fig. 5.1). Such extended shapes resisted being flipped over by currents, or being undermined

and buried by the scour and filling of mobile sediment. A/A' was therefore much < 1 (and is defined as < 0.75). Rudists of this morphotype were particularly associated with mobile lime sand to shelly rubble substrates, subjected to frequent winnowing and deflation, and with little or no sediment accumulation during their lives. Shells tended to be large, and of openly curved or radially splayed form (Fig. 5.2c).

The three morphotypes thus represent modes of growth suitable for different kinds of substrates. This does not mean that their distributions were mutually exclusive: a given facies could offer a mosaic of substrate-based microhabitats, allowing mixed assemblages. Morphotypes could also vary within some species and during ontogeny (juveniles were invariably attached clingers). The fabrics of rudist assemblages will be discussed below.

Before leaving morphotypes, we can now briefly examine the hypothesis of competitive displacement of corals from reefs. Central to the argument of Kauffman & Johnson (1988) is the idea that rudists evolved features convergent with those of framework-building corals, which brought them into competition with them. From a survey of several hundred species, several aspects of morphology were claimed to show 'trends' of convergence with 'reef-adaptive forms'. These 'trends' included: a progressive uncoiling of the valves — with reduction of the free valve — allowing greater packing; exaggeration of the external ornament; the evolution of stronger more predominantly aragonitic shells; and increasingly rapid erect growth (with more porous shell fabrics).

These supposed 'trends' are most closely linked with features of the elevator morphotype (cf. Fig. 5.2a). However, increased ornamentation was more characteristic in certain clingers (e.g. Fig. 5.2b—8, 9), whereas predominantly aragonitic shells were most characteristic of certain groups of recumbents (Masse & Philip, 1986). Kauffman's & Johnson's (1988) notion of the most 'reef-adapted' rudists is thus a chimaera, abstracted from the characteristics of many autecologically contrasting forms. Net changes in the frequencies of different morphological traits, abstracted from many rudist species, have been confused with evolutionary changes within lineages. To equate the two kinds of pattern presupposes that all rudist evolution was directed towards a single ecological 'goal', 'reef-adaptive morphology', whatever that may mean. This is an unacceptably teleological notion which ignores the manifest autecological diversity of rudists.

A second hypothesis advocated by Kauffman & Johnson (1988) to support their 'reef-building' model of rudist ecology is that of a widespread possession among them of symbiotic zooxanthellae. This popular idea was reviewed by Cowen (1983). It was argued, by analogy with hermatypic corals, that microbial photosynthesis could have boosted the rudists' rates of shell growth. It seems, however, that the hypothesis has won favour more by constant repetition than because of supporting evidence. Only a few rudist species show morphological adaptations for exposure of mantle tissue to the light, and these stand apart from other rudists because of their 'abnormal' morphologies (Skelton & Wright, 1987).

The symbiosis hypothesis seems particularly improbable for the most specialized of the elevators,

the hippuritids, with the exception of *Torreites* (Skelton & Wright, 1987). The pore and canal system in the hippuritid left valve is well suited for the circulation of feeding currents (Skelton, 1976), but shows no special modification for maximizing exposure of tissue to the light. Moreover, hippuritid thickets in the Santonian of NE Spain and SE France evidently thrived in relatively quiet, sometimes turbid waters, probably subject to nutrient fluxes (Gili, 1984; Grosheny & Philip, 1989). Indeed, some hippuritids persisted even in perideltaic lignite-rich marls (Mennessier, 1949). Grosheny & Philip (1989) compare their example, from La Cadiére in Provence, with the mud-rich 'oyster-reefs' of the North American Atlantic and Gulf coasts. Such nutrient-rich settings tend to be inimical to the development of hermatypic corals today (Hallock & Schlager, 1986).

It is clear, then, that the autecological attributes of rudists (reproduction, substrate relations, and nutrition) differed from those of reef-building hermatypic corals, and so their patterns of environmental distribution and response to disturbances can likewise be expected to have differed.

Other biota associated with rudist formations

A striking feature of Cretaceous bioherms and biostromes from the Aptian onwards is the overwhelming prevalence of rudists. In certain settings, corals, stromatoporoids, algae and problematical encrusting taxa made an important contribution (Masse & Philip, 1981; Scott et al., 1990). Nevertheless, compared with the Jurassic, hermatypic corals were depleted. Even in Barremian times, coral/stromatoporoid frameworks were largely restricted to outer-shelf or platform margins (Masse & Philip, 1981), and thereafter corals and stromatoporoids were generally rare in high current energy environments and grew preferentially in slightly deeper open marine settings (Scott, 1988, 1990; Scott et al., 1990). Calcareous algae were also of relatively low diversity in the Cretaceous, compared with the earlier Mesozoic, but red algae occurred in Upper Cretaceous rudist formations (Masse & Philip, 1981), and algal stromatolitic crusts accompanied laminar corals in the low energy zones of some Hauterivian to Albian examples (Scott, 1988, 1990). Encrusting problematica, including the *Lithocodium/Bacinella* organism and *Cayeuxia*, were locally common associates throughout the Cretaceous (Masse, 1979a; Simmons & Hart, 1987;

Camoin *et al.*, 1988).

Typical associates of the rudists also included other gregarious macrofauna, such as the sessile chondrodontid bivalves, which formed locally dense associations on Hauterivian to Cenomanian carbonate platforms, and the often large, free-living nerineid and acteonellid gastropods. Accompanying these were many facies-restricted benthic Foraminifera, including several 'giant' forms, such as the mid-Cretaceous orbitolinids and alveolinids, and the Campanian/Maastrichtian orbitoids. Examples of this fauna are illustrated in Carbone & Sirna (1981), Masse & Philip (1981), Simmons & Hart (1987), Scott (1990), Skelton *et al.* (1990), and by several papers in Carulli *et al.* (1989), from among an extensive literature.

Depositional fabrics of rudist formations

Rudists did not bud or branch in the manner of corals and sponges and thus could not themselves construct continuous branching reef frames. Moreover, algae and associated encrusters, though present in some rudist-dominated fabrics, typically grew as relatively thin surface crusts rather than pore-bridging crusts. Thus rigid organic framework development is rare in rudist accumulations. The rudists were, rather, largely occupiers of sedimentary substrates as is typical for aclonal benthic invertebrates (Jackson, 1985). Only a minority exploited hard substrates (see below), in contrast to the hermatypic corals and other colonial clonal forms.

Elevator rudists, especially certain hippuritids and radiolitids, formed the most striking of all in-place rudist accumulations. In these organ pipe-like congregations, elongate rudists grew vertically or inclined, forming sub-parallel fabrics (Figs 5.2a—6 and 5.3a). Typically, only one or a few species generated the fabric. Although individual shells may be a metre or more in length, it is likely that only the upper few centimetres projected above the substrate in life. Basal attachments are frequently small, and there may be surprisingly little contact between neighbouring individuals (Skelton & Gili, 1991), although examples of densely packed associations are known (e.g. Höfling, 1985). Usually the matrix sediment was thus an important means of support. These rudist 'thickets' were most common in relatively low current energy, shallow platform interior settings, where they formed extensive tabular bodies (Masse & Philip, 1981; Scott *et al.*, 1990). In contrast, those rudist fabrics that formed

in high-energy regimes tended to be associated with lime-sand to shell-rubble substrates. Recumbents (e.g. robust caprinids) dominated on current-swept surfaces. Even in the densest congregations, where considerable mutual shell contact developed, the horizontal growth adjustments of the shells precluded significant mutual attachment, and so rigid framework development was absent (Fig. 5.3b). This lack of framework meant that the growth fabrics were prone to storm disruption. Consequently only small lenses of *in situ* representatives of the living association are preserved in the rock record.

Between these end-member associations, dominated respectively by elevator and recumbent rudists, lies a variety of mixed associations. On grainy shelf margins localized but sometimes fairly diverse clusters of clingers (cf. Fig. 5.2b) and robust elevators (cf. Fig. 5.2a—5) occupied sites of intermittent sediment accumulation (Fig. 5.3c). Again, the lack of binding of these allowed their frequent disturbance by storms. Often they are preserved as para-autochthonous accumulations of floatstone or rudstone, with intact and broken shells (perhaps including those of washed-in recumbents, too) (Fig. 5.3d). Stable substrates in open water with little sediment accumulation, in contrast, allowed the localized development of boundstone fabrics comprising binding organisms with clinger and solitary elevator rudists (Fig. 5.3e and see also the Santonian 'complex buildups' described in Scott *et al.*, 1990).

The sediment-dominated character of most rudist formations explains our reluctance to use the term 'reef' to describe them (see also Gili *et al.*, 1990). Even if the sort of generalized definition of the term that only requires the joint presence of framework and of origin topographical relief is employed, very few rudist formations qualify. Although elevator thickets might be considered sediment-trapping 'frameworks', the modest original topographical relief of these bodies rules them out as 'reefs'. On the other hand, although recumbent and/or mixed associations may form a significant part of shelf and platform margins, their generally para-autochthonous fabric, and trivial development of framework, excludes them from being considered reefs as well. Rare examples with both attributes (e.g. Camoin *et al.*, 1988) are known but are of only small scale. Nowhere is there any documented evidence for the rudists having built large-scale, organic framework-supported edifices

(a)

(b)

(c)

(d)

(e)

Fig. 5.3 Examples of rudist-dominated fabrics. (a) Thicket of hippuritid elevators (cf. Fig. 5.2a—6). Bedding is vertical, top to left (Santonian of Les Collades de Basturs, central southern Pyrenees; topmost inner platform thicket of Gili, 1984; Scott *et al.*, 1990). (b) Stacked recumbent caprinids (cf. Fig. 5.2c—11). Upper surface and, in foreground, vertical section of isolated mound in volcanic sequence (?Lower Cenomanian of Paso del Rio Armería, Periquillos, Colima Province, Mexico; see Huffington, 1981). (c) Bouquet of clinger to elevator radiolitids (cf. Figs.5.2a—5, b—8) in rudist floatstone, in vertical section (Upper Campanian/?Lower Maastrichtian of Poggiardo, Salento Peninsula, SE Italy; see Cestari & Sirna in Carulli *et al.*, 1989). (d) Para-autochthonous rudist floatstone with various radiolitids and hippuritids, in vertical section (as in c). (e) Boundstone, with clinger radiolitid and *Lithocodium* binding (pale crust around rudist), in vertical section (?Cenomanian of Wadi Laasi, Jebel Nakhl, near Muscat, Oman).

comparable with those of some wave-resistant coral reefs.

Despite the variability of rudist depositional fabrics, certain common features should be emphasized because of their implications for diagenesis. Due to the absence of a well-developed frame, framework cavities are limited. Primary intergranular porosity would have been present in coarse grain-supported sediments (Fig. 5.4d) and large intraskeletal cavities were present within the rudists themselves (Fig. 5.4a). The intraskeletal cavities are important in that they frequently preserve protracted cementation sequences (Fig. 5.4b, c). Rudists are thus often useful keys for the diagenetic history of sediments which would otherwise have failed to preserve full cementation sequences. In addition, in many rudist formations, extensive secondary mouldic porosity was caused by dissolution of the aragonitic parts of rudist shells. The outer calcitic shell layers were usually unaffected by dissolution and, where strong enough, acted as rigid casings allowing the preservation of fragile internal collapse features (Fig. 5.4b). Thus, although diagenetic modifications of matrix textures are usually somewhat cryptic, the rudists themselves offer great potential for secondary porosity enhancement and for diagenetic studies in general.

Physiography of rudist formations

Because rudists and the character of rudist formations vary so greatly there is a need for a conceptual framework within which to compare individual examples. Using platform setting as a basis of comparison, we have drawn on numerous examples to make generalizations concerning the nature of rudist formations in relation to setting (Fig. 5.5). Examples from each setting are presented, and, while there may be considerable variation and exceptions, the scheme nevertheless represents a step towards improving the predictability of rudist formations.

Steep margin complexes

Steep margins are defined as those which had slope angles in excess of 10°, and faced oceanic or fault-bounded basins. Margin growth was usually strongly aggradational, with slopes frequently accentuated by syn-depositional faulting. Albian to Cenomanian examples in north-eastern Mexico (see Scott, 1990) and north-western Israel (Ross, 1992)

show vertically continuous sections of marginal facies several hundred metres thick (Fig. 5.6). More common are smaller-scale steep intraplatform margins, generated by extensional faulting (e.g. Lower and Upper Cretaceous examples in SE Italy, described by Borgomano & Philip, 1989; Masse & Luperto-Sinni, 1989; and Pieri & Laviano, 1989). Lateral facies variations are commonly abrupt (Fig. 5.6). Grainy slope or toe-of-slope deposits may extend some kilometres basinward of the margin. On the upper slopes of both of the mid-Cretaceous examples shown in Fig. 5.6, radiolitid elevators to clingers locally formed dense clusters, frequently preserved in place, embedded in shelly packstones to grainstones partly derived from shallower levels. Grainstones, with abundant shell rubble, accumulated in thick to massive beds on the shallow crest of the margin. Chondrodonts and nerineid gastropods dominate this facies in Israel (Fig. 5.6a). In the El Abra platform margins (Fig. 5.6b), rudist lenses of modest dimensions (tens of metres wide and < 10 m thick) are intercalated with the grainstones. Fabrics within the lenses vary: discrete associations of large recumbent caprinids in shelly grainstones, and slender elevator caprinids in muddier sediments probably reflect transient fluctuations in current regime (Collins, 1988). On the Campanian/ Maastrichtian faulted platform margins of SE Italy, a large recumbent rudist ('*Sabinia*') is common, though frequently transported downslope, along with lithified blocks containing other rudists (Borgomano & Philip, 1989; Pieri & Laviano, 1989). The latter provide important evidence of early cementation of some of the platform-crest sediments.

At the back of the grainstone belt there may be extremely coarse grainstones with multiple disconformity horizons. At Nahal Ha'Mearot, in Israel, these have been shown to indicate repeated emergence, with grainstones forming shelf-margin sand cays set back from the outer margin (Fig. 5.6a; Ross, 1992). Yet further from the margin, these pass to muddy inner-platform facies (Fig. 5.6).

The absence of any coherent organic framework-supported edifice, forming a projecting wave-resistant outer rim, on these and other steep margins dominated by rudists should be noted. Rather, they were 'round-shouldered' carbonate bank complexes, frequently shedding large quantities of grainy bioclastic sediment downslope. The rudists which infested the upper slopes and crests, both helped to stabilize, and, particularly in the late

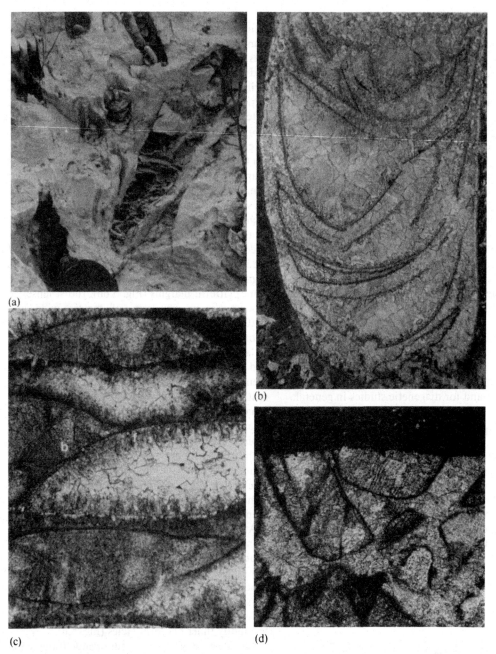

Fig. 5.4 Diagenesis of rudist shells. (a) Elevator rudists showing preserved calcitic outer shell layer and leached (originally aragonitic) tabulate inner shell, with intraskeletal cavities (Upper Campanian of SE Italy). (b) Photomicrograph of oblique section of elevator (hippuritid) rudist showing calcitic outer shell (dark) and collapsed cement crusts of leached tabulae; primary and secondary porosity now occluded by calcite spar (Santonian of Sardinia). Width of frame is approxmately 4 mm. (c) Local inhomogeneities in fringe (f) and botryoidal (b) early cements on the reticulate tabulae of a hippuritid rudist (Santonian of south central Pyrenees, Spain). Width of frame is approximately 4 mm. (d) Grainstone matrix of a rudist biostrome, showing early (marine) cementation and erosional surface overlain by lime mudstone (Cenomanian of Istria, NW Yugoslavia). Width of frame is approximately 2 mm.

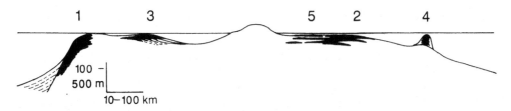

Fig. 5.5 Platform settings of rudist formations (shown in black): (1) steep margin complexes; (2) low-angle open shelf margin complexes; (3) inner shelf basin prograding margin complexes; (4) isolated build-ups; and (5) inner shelf and platform biostromes.

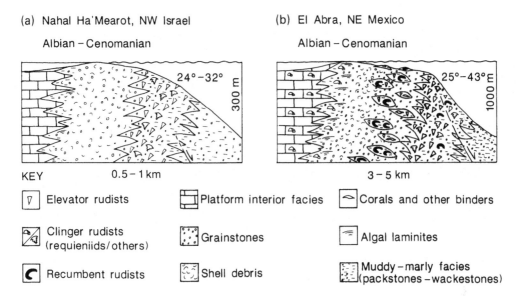

Fig. 5.6 Examples of steep margin rudist complexes, in diagrammatic section. (a) Nahal Ha'Mearot, NW Israel (Albian–Cenomanian: see Ross, 1992). (b) El Abra platform margins of NE Mexico (Upper Albian–Lower Cenomanian: see Scott, 1990).

Cretaceous, contributed abundant bioclasts to these sediments. Moreover, early cementation and re-working of some of these sometimes supplied lithified blocks to the slope deposits.

Low-angle open shelf margin complexes

Margins in this category are those with original slopes of less than 10° (often less than 3°), which faced open marine (including oceanic) basins. Depositional geometries are commonly progradational (Fig. 5.7) and may exhibit multiple shoaling-up sequences up to tens of metres in thickness (e.g. Bebout *et al.*, 1981; Scott *et al.*, 1990; see Fig. 5.7b). Corals and sponges (especially stromatoporoids) may be relatively common in the lower beds of the

progradational sequences, but rudists dominate the upper beds (Fig. 5.7). This vertical sequence reflects a bathymetric faunal zonation, with the corals and sponges occupying deeper slope settings, and the rudists, shallower, shelf-top environments (Scott, 1988; Scott *et al.*, 1990).

The slope deposits are mainly of low-current energy character: packstones and wackestones predominate. Extensive lenses of boundstone, comprising encrusting corals, stromatoporoids and algae, with subsidiary rudists, developed downslope from the crest in some early examples, such as the Aptian upper Sligo shelf margin (Bebout *et al.*, 1981) and the middle Albian Stuart City margin (Fig. 5.7a), both in the subsurface of southern Texas. Coral/rudist associations also persisted,

(a) Stuart City trend, S Texas

Albian

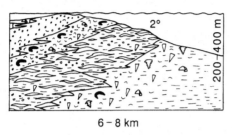

(b) Les Collades de Basturs, NE Spain

Santonian

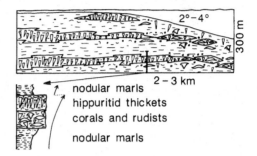

Fig. 5.7 Examples of low-angle open shelf margin complexes in diagrammatic section. (a) Stuart City trend, subsurface of S. Texas (Middle Albian: see Scott, 1990). (b) Les Collades de Basturs, south central Pyrenees, Spain (Santonian: see Scott *et al.*, 1990). Diagrammatic log, below, illustrates progradation of inner platform thickets over outer platform coral–rudist associations. Key symbols as in Fig. 5.6.

though with a less extensive development of bound-stone fabrics, on the flanks of many Upper Cretaceous low-angle margins (e.g. Carbone & Sirna, 1981; Masse & Philip, 1981; Scott *et al.*, 1990).

Outer margin facies varied according to palaeogeographical situation. On more hydrodynamically active, ocean-facing margins, such as the Sligo and Stuart City trends (Fig. 5.7a), grainstones and packstones accumulated, locally infested by recumbent (caprinid) and clinger (requieniid) rudists, together with scattered corals. On less exposed margins, such as the Santonian Les Collades de Basturs margin of the south-central Spanish Pyrenees, small 'complex' lenses, with mainly solitary large elevator and clinger rudists (hippuritids and radiolitids) and massive corals, were interdigitated with marly packstones and nodular marls (Fig. 5.7b, right side). Behind these marginal complexes tabular hippuritid thickets developed in the quiet waters of the platform top (Fig. 5.7b left side, and see below).

Inner-shelf basin prograding margin complexes

Inner-shelf basins were relatively deep-water depressions within extensive carbonate shelves. Such basins were typically a few tens to a few hundreds of metres deep and had lateral extents of up to several hundred kilometres. At the shallow margins of these basins, high current energy conditions often prevailed, promoting the development of grainstones (Fig. 5.8). These commonly incorporated biostromes of recumbent and clinger rudists, while downslope, muddy build-ups with elevators, some-

Mishrif margin, Abu Dhabi (offshore)

Cenomanian

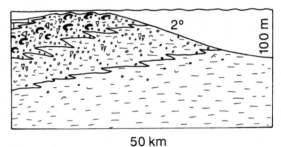

50 km

Fig. 5.8 Example of inner shelf basin prograding margin complex in diagrammatic section: Mishrif platform margin, offshore subsurface of Abu Dhabi, United Arab Emirates (Cenomanian: see Burchette & Britton, 1985). Key symbols as in Fig. 5.6.

times developed below wave-base. Corals were generally rarer than on ocean-facing low-angle margins, probably because of the fluctuating water chemistry of these more or less enclosed marine basins.

Perhaps the best known examples of inner-shelf basin margin rudist formations are those rimming the deep water Shilaif Arm of the Rub al Khali Basin of the Arabian Peninsula. These occur in the Aptian Shuaiba Formation and the Albian to Cenomanian Mauddud and Mishrif Formations (Alsharhan & Nairn, 1988). In these, extensive centripetal progradation produced broad, tabular carbonate platforms, with gently sloping margins encircling the basin. For the Mishrif platform (Fig. 5.8), Burchette & Britton (1985) inferred a very low gradient margin (devoid of slump features), 'over which shoal-derived sediment dispersed and fined basinwards in response to ebb-tidal or ebbing storm-surge currents'. The shoal facies at the margin crest comprises coarse bioclastic packstones and grainstones with largely para-autochthonous rudist biostromes. Although Burchette & Britton (1985) assert that small radiolitids dominate here, it is clear from their own core photographs (Burchette & Britton, 1985, Fig. 9), as well as from study of other Mishrif material by PWS, that recumbent caprinids and ichthyosarcolitids are equally, if not more, abundant. The recumbents, which were originally largely aragonitic, have frequently been leached, however, leaving somewhat non-descript vugs which are less easily recognized than the better preserved (largely calcitic) radiolitids. Also common in this crest facies are repeated coarsening-upwards progradational units, each of a few metres thickness, which probably represent migrating banks. Downslope, muddy bioclastic sediments accumulated (Fig. 5.8). On the Shuaiba margin, downslope muddy build-ups also developed, with abundant *in situ* elevator caprotinids (Frost *et al.*, 1983).

A very similar facies geometry has recently been documented for a prograding inner-shelf basin platform of Campanian age on Brač Island, western Yugoslavia, by Gušić & Jelaska (1990). Here, the grainy crest facies, with scattered large clinger and stellate recumbent radiolitids (cf. Fig. 5.2c–12) and solitary elevator hippuritids built out as a tabular unit some 20 m thick over hemipelagic sediments. Small rudist lenses also developed on the muddy lower slopes, dominated by elevator and clinger radiolitids and hippuritids.

As in the other examples discussed, foraminiferal packstones to wackestones accumulated behind the shoal facies.

Isolated build-ups with rudists

Isolated build-ups are those which grew up locally from the sea-floor, and which were usually initiated on sea-floor highs (commonly biogenic mounds, structural highs, salt pillows or volcanic extrusions; Fig. 5.9). Many early Cretaceous examples commenced with stacked coral/mud mounds. Fabrics

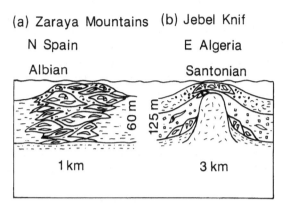

(a) Zaraya Mountains   (b) Jebel Knif

N Spain   E Algeria

Albian   Santonian

Fig. 5.9 Examples of isolated coral–rudist build-ups in diagrammatic section. (a) Zaraya Mountains, N. Spain (Upper Aptian: see Fernandez-Mendiola & Garcia-Mondejar, 1989). (b) Jebel Knif, Khenchela, E. Algeria (Santonian: see Camoin *et al.*, 1990). Small build-ups flank and cap an exposed diapir (centre) of Triassic evaporites and associated sediments. Clasts of the latter (squares) dominate the flanking facies. Key symbols otherwise as in Fig. 5.6.

varied from floatstone with wackestone matrix to bafflestone or boundstone (Masse & Philip, 1981; Fernandez-Mendiola & Garcia-Mondejar, 1989). These are considered to have built up below wave-base. Binding organisms such as *Lithocodium/Bacinella* nevertheless reinforced the mounds, and allowed their upward growth. In some late Cretaceous mounds, the preservation of inferred microbial textures in steeply dipping (up to 40°), flanking mud- and grain-supported deposits has lead to suggestions that microbial action may have helped stabilize low-energy isolated rudist build-ups (Camoin *et al.*, 1988).

With shoaling into wave-base, grainstones developed on top and were shed around the flanks. Here (as in the sequences discussed above) rudists proliferated in the crest facies, e.g. clinging requieniids (Fig. 5.9a), and/or recumbent caprinids, perhaps with chondrodontids. The development of marginal banks led, in some cases to an atoll-like profile (e.g. the James Atoll 'Reef', in Texas; Achauer, 1983). Examples also occur on seamounts in the Pacific (e.g. Konishi, 1985). In less hydrodynamically energetic settings, mud-supported coralgal boundstones and elevator rudist clusters developed (e.g. the flanking lenses of Camoin *et al.*, 1990; Fig. 5.9b).

Inner-shelf and platform rudist biostromes

Carbonate shelf and platform interiors were the sites of formation of laterally extensive rudist biostromes (Masse & Philip, 1981). Rarely more than a metre or two thick, these tabular bodies had little original topographical expression, and usually passed laterally into surrounding sediments with no discrete flanking deposits. Their diversity is typically low (a single species may be dominant), and the matrix is generally lime-muddy to marly wackestone to packstone.

In the Lower Cretaceous, these biostromes mainly comprise para-authochthonous floatstones of requieniid clingers, or sometimes in-place clusters or small thickets of primitive elevators such as *Monopleura* and *Agriopleura* (e.g. Masse, 1979b; Urgonian of SE France), and, later, radiolitids and caprinids (e.g. Scott, 1990: Albian of Texas).

In the Upper Cretaceous, elevator radiolitids and, from the Turonian onwards, hippuritids developed thickets or para-autochthonous floatstones, according to local circumstances of sediment deposition. Thickets commonly only comprise one or a few generations, but can nevertheless extend laterally as tabular units of hundreds of metres across (e.g. Grosheny & Philip, 1989). Similar thickets developed behind the shelf edge at Les Collades de Basturs (Fig. 5.7b, left side; Gili, 1984; Scott *et al.*, 1990). Their relatively low diversity and sporadic occurrence as discrete units, representing only one or a few generations of growth, suggest occasional rapid opportunistic development. In some cases, conditions were evidently inimical to the growth of hermatypic corals (as noted earlier).

Some palaeogeographical differentiation of these Upper Cretaceous thickets is evident: the broad carbonate platforms and shelves of the low-latitude southern Tethys were commonly exclusively occupied by elevator radiolitids (e.g. the Coniacian/Santonian Altamura Formation of Apulia, Pieri & Laviano, 1989; and the Maastrichtian Sumartin Formation of Brač Island, Gušić & Jelaska, 1990). Hippuritids were included in more open settings in some areas (e.g. Sirna & Cestari, 1989), but these became dominant, by contrast, on many of the marly northern Tethyan shelves (e.g. Masse & Philip, 1981; Höfling, 1985; Scott *et al.*, 1990), which were situated at higher palaeolatitudes. While some radiolitids were capable of prolonged hermetic closure of the valves (Skelton *et al.*, 1990),

hippuritids, with their open system of pores and canals (Skelton, 1976) were probably incapable of entirely excluding the ambient water. It is tempting, therefore, to suggest a climatic/environmental control (perhaps involving salinity fluctuation) to explain these distributional differences. However, further work on this question is needed.

The stratigraphical history of Cretaceous rudist formations

The rudists had several phases of radiation during the Cretaceous, punctuated by crises of mass extinction (Masse & Philip, 1986; Skelton, 1991; Skelton & Gili, 1991). These evolutionary changes, and associated eustatic, tectonic and oceanographic events caused the nature of rudist formations to change during this period. In this section we identify four phases of development and three major crises in the history of rudist formations (Fig. 5.10).

Phase 1: early Cretaceous to the mid-Aptian — initial diversification of rudist formations

The earliest rudists (late Jurassic) were clingers, spirally overgrowing the substrate with the flattened anterior face of the AV (Skelton, 1978). The requieniids continued this habit throughout the Cretaceous (Fig. 5.2b—7). At the close of the Jurassic, the first 'uncoiled' (caprotinid) rudists appeared, and adaptive radiation ensued, with examples of all three morphotypes evolving (Skelton, 1985).

Throughout this initial phase, largely para-autochthonous biostromes of clingers and elevators developed on usually sheltered platform interiors (Masse, 1979b; Masse & Philip, 1981). Recumbents (notably the caprinids, cf. Fig. 5.2c—10) appeared in the early Aptian (Fig. 5.10; Skelton, 1991), and these flourished on more current-swept lime-sand substrates (e.g. in Apulia, Luperto Sinni & Masse, 1982; and on the Sligo margin of southern Texas, Bebout *et al.*, 1981).

Coral/stromatoporoid/algal build-ups and biostromes, by contrast, were already largely confined to outer platform and shelf settings, replaced in areas of high-sediment mobility by migrating carbonate sand bodies (Masse & Philip, 1981). There was therefore little overlap in the biotopes occupied by the rudists and the hermatypic corals — a clear weakness in the competitive displacement hypothesis of Kauffman & Johnson (1988). Much more

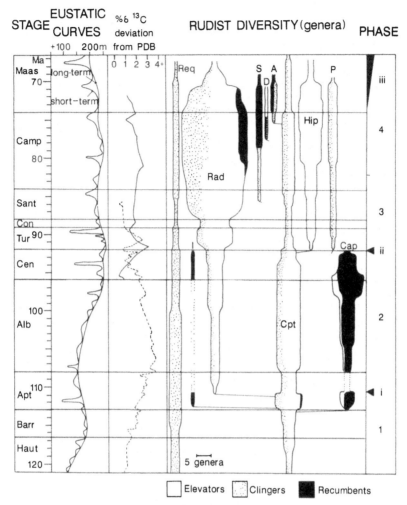

Fig. 5.10 Stratigraphical history of rudist morphotypes in the Cretaceous (Hauterivian–Maastrichtian). Absolute time scale and inferred eustatic curves (metres above present sea-level) from Haq *et al.* (1987). Deviations of carbon isotope ratios from PDB standard, in whole-rock samples of pelagic limestones from Scholle & Arthur (1980): ——, Norfolk, UK; ----, Peregrina Canyon, USA. Standing generic diversity of rudists modified from Skelton & Gili (1991) and Skelton (1991): note that diversity values are approximate, because of the need for further systematic and stratigraphical revisions, and also that diversity data have been assembled on the basis of chronostratigraphical distribution, not on that of absolute age. Req, Requieniidae; S, *Sabinia* species group; D, *Dictyoptychus* group; A, Antillocaprinidae; P, Plagioptychidae; Rad, Radiolitidae; Hip, Hippuritidae; I, Ichthyosarcolitidae; Cap, Caprinidae, Cpt, Caprotinidae. Phases in the history of rudist formations (1–4), and crises of mass extinction (i–iii) are discussed in the text. Note the lack of clear correlation between the short-term changes in the interpreted eustatic curve and rudist history, but the approximate coincidence of inferred productivity peaks, eustatic highstand and the demise of recumbent rudists around the Cenomanian–Turonian boundary.

plausible is a model of the kind advocated by Scott (1988), whereby coral reefs became restricted by environmental perturbations (such as fluctuating nutrient levels), with rudists independently diversifying in a range of platform and shelf facies. Scott's (1988) assumption of widespread symbiosis of zooxanthellae in rudists (perhaps allied with his allegiance to the notion of rudists as reef-builders) is doubtful, however, for the reasons given earlier. It may indeed have been the case that the rudists' *lack* of symbionts made them less sensitive to fluctuations in nutrient levels than the hermatypic corals.

Another speculative possibility is that patterns of carbonate sediment production and transport on platforms and shelves in the early Cretaceous led to a predominance of mobile sand facies, especially on their margins. These would have been inimical to the establishment of coral (or any other kind of) frameworks, but favourable for the spread of the various sediment-adapted rudist assemblages mentioned above. Such rudist facies are particularly well-developed in the early Aptian (Masse & Philip, 1986). The notion of rudists invading reefs might perhaps be replaced by that of reefs declining and rudist-infested sedimentary facies expanding. Much further work on the facies geometry of Lower Cretaceous carbonate platforms and shelves is needed, however, to clarify the nature of these changes.

Crisis 1: the mid-Aptian mass extinction

Lower Aptian rudist limestones terminate virtually everywhere with an abrupt facies change. Although palaeobathymetric change is usually evident, there is not a universal signature of shallowing or deepening, and so the possible roles of eustasy, tectonism, climate and/or changes in oceanic circulation or chemistry have not been clearly resolved. Whatever the cause, this level marks an important extinction event, affecting calcareous algae and benthic Foraminifera, as well as rudists (Masse, 1989). Recumbents, in particular, disappeared (Fig. 5.10), though the later recurrence of some of the genera in the Albian/Cenomanian implies the existence of some, as yet, undiscovered refuges (Skelton, 1991).

Phase 2: late Aptian to Cenomanian — first acme of rudist formations

Recovery from the mid-Aptian event(s) was slow: rudist formations of definite late Aptian age are, as yet, unrecorded in the New World. They are present in the Old World, although only clinger and elevator rudists (including early radiolitids) are involved (Fig. 5.10; e.g. Masse & Philip, 1981; Fernandez-Mendiola & Garcia-Mondejar, 1989). Thereafter, the history of rudist formations in the New and Old Worlds shows a curious contrast. The Albian to early Cenomanian saw the extensive development of carbonate shelves and platforms in Mexico and around the Gulf coast, and rudist associations of every type flourished on them (reviewed in Scott, 1990; and above). In particular, endemic recumbent caprinids became diverse and abundant (Skelton, 1991). In the Old World, in contrast, Albian

rudist formations still only contained clingers and elevators (e.g. Scott et al., 1990; Ross, 1992). Here, the main expansion of recumbent-dominated high-current energy facies (coarse rudstones to grainstone–floatstones) was in the Cenomanian (Masse & Philip, 1986; Skelton, 1991; and see other examples discussed above). The reasons for this palaeogeographical distinction remain unresolved.

Coral/stromatoporoid/algal reefs persisted, though these were generally in downslope situations (Scott, 1990, and above).

Crisis 2: the Cenomanian/Turonian boundary — platform drowning and mass extinction

Sharp facies changes mark the end of this second phase of development, which culminated in widespread platform drowning, associated with a eustatic peak in the early Turonian (Hancock & Kauffman, 1979; Fig. 5.10). Sea-level rise was accompanied by a major mass extinction of platform biota, especially rudists (Masse & Philip, 1986). Again, recumbents disappeared (Fig. 5.10), and this time the families involved (caprinids and ichthyosarcolitids) became entirely extinct (Skelton, 1991). Schlager (1981) has argued, however, that long-term rates of sea-level rise are slow compared with carbonate platform accretion rates and alone cannot explain platform drowning. The correlation between platform drowning and anoxic events led Hallock & Schlager (1986) to suggest that, during anoxic events, nutrient floods from mid-oceanic overturn damaged autotrophic carbonate benthos and reduced carbonate platform growth rates, thus allowing platform drowning. The major peak in δ^{13}C recorded in pelagic sediments during the early Turonian by Scholle & Arthur (1980) may be interpreted as supportive evidence for high oceanic nutrient levels at this time (Fig. 5.10). However, tectonic influences on relative bathymetry are àlso invoked by some authors (e.g. Philip et al., 1989). As with the earlier crisis, more work on the correlation and nature of events at this time is needed in order to resolve causes and effects.

Phase 3: Turonian to early Campanian — partial re-establishment of rudist formations

During the Turonian to Santonian, there was a radiative evolution of the surviving rudist faunas. Of particular importance was the evolution of the hippuritids which were to remain important in rudist formations through to the Maastrichtian

(Fig. 5.10). In Turonian to Santonian biostromes, elevator rudists (radiolitids and hippuritids; see Fig. 5.2a) predominated (Masse & Philip, 1981) and usually formed thickets in relatively quiet waters. More hydrodynamically active environments at this time were apparently more often the sites of oolitic and bioclastic grainstone accumulation than rudist biostrome development (e.g. Carbone & Sirna, 1981).

The lack of recumbent taxa at this time (Fig. 5.10) does not satisfactorily explain the rarity of high current energy rudist facies: the rapid polyphyletic evolution of recumbent taxa in the succeeding phase implies that such forms could evolve rapidly when conditions were favourable (Skelton, 1991). It is therefore likely that there were environmental constraints on the development of high-energy rudist facies during this approximately 10 m.y. period, but what these might have been remains unclear.

Phase 4: mid-Campanian to early Maastrichtian — the second acme of rudist formations

The fourth phase represents a continuation of the third, though with whatever had been the constraint on the evolution of recumbents now gone. During this interval a full spectrum of rudist formations again developed (see above), and, in particular, high current energy platform- and shelf-margin rudist formations became common. New taxa of large recumbent rudists became prominent in these (Fig. 5.10), including antillocaprinids in the Caribbean and southern Mexico, and the polyphyletic genus 'Sabinia', together with certain bizarre stellate radiolitids (e.g. Fig. 5.2c–12) in the Old World (Skelton, 1991). Problems of correlation have caused much confusion over the dating of these Old World accumulations. Though many examples were earlier considered to be Maastrichtian, and certain rudist genera thereby gained false reputations as 'Maastrichtian markers', a mid- to late Campanian age now seems more likely for most (e.g. Gušić & Jelaska, 1990; and strontium isotope correlations of Swinburne, 1990), with some creeping into the early Maastrichtian (e.g. Skelton et al., 1990).

Crisis 3: the Maastrichtian — final demise of the rudists

Unlike the earlier two crises, this final extinction episode seems to have been a longer-term decline. The revised datings mentioned above suggest that rudist diversity dwindled through the early Maastrichtian and was much depleted in the late Maastrichtian (Fig. 5.10). Progressive loss of habitats with the diminution of carbonate platforms and shelves at that time (involving both drowning of some examples and emergence of others through a combination of tectonism and eustatic changes) is the most probable explanation (Swinburne, 1990). Scenarios incorporating the demise of the rudists in the terminal Maastrichtian extinction event ('the K-T Boundary Event'), or even declining only from 'mid-Maastrichtian' times are not supported by the currently available evidence. Upper Maastrichtian rudist formations appear to be limited to small-scale, inner-platform biostromes with elevators and clingers (e.g. Bilotte, 1985; Drobne et al., 1988).

Hydrocarbon prospectivity of rudist formations

Rudist formations and associated sediments are important oil and gas reservoirs in the Middle East, around the Gulf of Mexico and to a lesser extent in the Mediterranean realm of Tethys. The hydrocarbon potential of rudist formations is primarily controlled by type and setting, by stratigraphical context and by diagenetic effects.

At steep shelf margins the aggradational nature of the facies, their grainy textures and proximity to basinal source rocks, make these formations attractive prospects. Margins bordering the ancestral Gulf of Mexico, for example the Golden Lane trend in Mexico, prove this potential (Enos, 1977). An additional major contribution to hydrocarbon potential at steep shelf margins is made by the toe-of-slope deposits which interdigitate with and are sealed by basinal source rocks. The Poza Rica field, Mexico's largest oil field for 50 years, is from this setting (Enos, 1988). By comparison, low-angle open margins are generally less attractive prospects. Margin facies are less vertically continuous and have a greater proportion of mud-supported fabrics. Moreover, slope deposits are of a 'lower energy' character and include only rare grainstones. The Stuart City trend of Texas, which occurs at a low-angle shelf margin was explored as an analogue for the Golden Lane and Poza Rica trends of Texas (Bebout & Loucks, 1974). However, as a result of poor primary porosity development and with a lack of karstic porosity, only small reserves were discovered.

Some rudist-dominated prograding marginal facies of inner-shelf basins contain significant hydrocarbon reserves, particularly in the Middle East.

The Aptian Shuaiba Formation includes the giant Bu Hasa field of Abu Dhabi (Alsharhan, 1987), whilst the Albian to Cenomanian Mauddud and Mishrif Formations include the Fahud and Natih fields of Oman (Harris & Frost, 1984) and the Fateh field in Dubai. Factors contributing to prospectivity are similar to those for steep open margins (despite the paradoxically gentle slopes of the inner-shelf basin examples). These are a dominance of grainstones and proximity to organic-rich source rocks. The relative paucity of coral boundstones in both cases is an intriguing common factor.

The hydrocarbon potential of rudist build-ups on isolated highs is strongly influenced by whether or not they grew above wave-base. High current energy isolated build-ups have many of the attributes of steep margins and are proven hydrocarbon plays, e.g. the James Atoll 'Reef', in Texas (Achauer, 1983). By contrast, low current energy build-ups are less attractive, being dominated by low permeability mud-supported textures.

Beside platform setting, a second fundamental control on the prospectivity of rudist formations is their stratigraphical context. Although local stratigraphy is primarily influenced by regional basin development, certain events are of world-wide significance. The fact that the vast majority of proven reserves in rudist formations are of Albian to Cenomanian age demonstrates this. Complexes of this age were commonly deposited in high current energy conditions; the thick successions of stacked grainstones colonized by recumbent rudists, which are typical of margins of this age, form excellent reservoirs. Equally important are the mid-Cretaceous platform-drowning events which frequently resulted in the formation of regional seals and excellent source rocks. These combined factors of reservoir, source and seal development at least partly explain the bias of accumulation of hydrocarbon reserves in rudist formations of the Albian to Cenomanian interval.

After platform setting and stratigraphical context, the most crucial influence on the hydrocarbon potential of rudist facies is their diagenetic evolution. Rudists, particularly the recumbent forms which characterize high current energy facies, had thick aragonitic inner-shell layers. Thus whole shells and fragmented debris were prone to dissolution (Fig. 5.4). This secondary enhancement of porosity by the development of mouldic macro- and microporosity is a vital factor in the production performance of rudist facies. Dissolution of aragonitic components is generally attributed to the influence of meteoric water, either by emergence or via subsurface aquifer charge. However, under meteoric influence a fine balance exists between dissolution and cementation. Thus, although aragonite dissolution can be recognized as important, the presence of secondary porosity is difficult to predict even where meteoric influence is suspected.

Conclusions

In the Cretaceous, the rudist bivalves evolved elevator and recumbent morphotypes, joining existing clinger morphotypes (Fig. 5.1). These diverse forms were superbly adapted to colonization of a variety of sedimentary substrates, but poorly suited to framework development. Rudists did not bud or branch in the manner of colonial corals. Moreover, the upwardly growing elevators had only small basal attachments, and thus growth fabrics relied upon sediment support for stability, and were susceptible to disruption by storms.

The rudists proliferated on carbonate platform and shelf interiors in the early Cretaceous, with biostromes containing clingers and elevators. With the advent of recumbents in the early Aptian, the higher current energy lime-sandy facies of platform and shelf exteriors were also invaded. Thereafter, for the rest of the Cretaceous, rudists dominated a wide variety of facies on platform and shelf tops and on margins, commonly forming dense associations. The nature and stratigraphical distribution of these were influenced both by palaeogeography (Fig. 5.5) and by rudist evolution: the latter involved several phases of radiation, punctuated by mass extinctions in the mid-Aptian, around the Cenomanian/Turonian boundary, and running up to the end of the Maastrichtian (Fig. 5.10).

Scleractinian coral/'stromatoporoid' sponge/algal boundstones had, by the early Cretaceous, already become mainly limited to outer platform and shelf settings. They persisted, to a lesser extent, through the rest of the period, though largely limited from the Aptian onwards to downslope settings on open marine low-angle margins. In these they formed lenses in relatively low-current energy zones during episodes of limited sedimentation.

This complex pattern of changes in biota and facies is probably best explained in terms of the independent decline of coral reefs and spread of rudist formations. Possible influences include fluctuations in marine nutrient supply and changes in the facies geometry of Cretaceous carbonate shelves and platforms. Marked differences in autecology between rudists and hermatypic corals suggest that there was little (if any) competitive interaction between them, and the model of competitive displacement is rejected.

At steep margins bordering open oceans, on gentler slopes around the prograding margins of inner-shelf basins, and on shallow isolated highs, high current energy substrates were colonized by recumbent rudists. The resultant accumulations, together with associated slope grainstones, contain considerable hydrocarbon reserves around the world. By contrast, on low-angle open margins, and on deep-water isolated highs, elevator rudists formed mud-based clusters and thickets. In these settings rudists were also locally associated with corals, algae and sponges to form diverse fabrics tending towards boundstones. However, the dominance of mud-supported textures makes these build-ups less attractive as hydrocarbon prospects than the high current energy facies of steep and inner-shelf basin margins.

The age of a sequence is an important consideration for predicting formation type. During periods when recumbent rudists were diverse (early Aptian, Albian to Cenomanian and Campanian to Maastrichtian), high current energy facies dominated by these forms were developed. By contrast, after extinction episodes, sometimes accompanied by platform drowning, rudists were restricted to lower current energy settings and in these only elevators and clingers were present, sometimes associated with corals and algae.

Rudist formations were thus remarkably diverse, both in time and space. It is important to recognize this, and to develop a range of suitable models, based upon direct field observations, if they are to be properly understood, whether for the interpretation of palaeogeography or in the search for hydrocarbons. The uncritical application of simplistic 'reef' models, based upon Recent coral/algal reefs, has for too long impeded progress in the interpretation of rudist formations, and has yielded some frankly misleading conclusions. This paper is an attempt to break with that stultifying tradition.

Acknowledgements

D.J. Ross's doctoral research on rudist formations was funded by a NERC/CASE award with BP, held at University of Wales College of Cardiff, and supervised by Robert Riding and Trevor Burchette. P.W. Skelton acknowledges numerous helpful discussions on rudist palaeoecology with Eulalia Gili (Barcelona, Spain) and Jean-Pierre Masse (Marseille, France), although any errors of fact or interpretation herein are the responsibility of the authors alone. Their comments on this chapter as well as the helpful advice of two referees are also gratefully acknowledged. The manuscript was processed at the Open University by Janet Dryden, and the photographic figures prepared by John Taylor.

References

Achauer, C.W. (1983) Reef, lagoon and off-reef facies, James Atoll Reef (Lower Cretaceous), Fairway Field, Texas. In: *Carbonate Buildups — a Core Workshop* (Ed. P.M. Harris). Soc. Econ. Paleont. Miner. Core Workshop 4, 411–428.

Alsharhan, A.S. (1987) Geology and reservoir characteristics of carbonate buildup in giant Bu Hasa Oil Field, Abu Dhabi, United Arab Emirates. *Bull. Am. Ass. Petrol. Geol.* **71**, 1304–1318.

Alsharhan, A.S. & Nairn, A.E.M. (1988) A review of the Cretaceous formations in the Arabian Peninsula and Gulf: Part II. Mid-Cretaceous (Wasia Group) Stratigraphy and Paleogeography. *J. Petrol. Geol.* **11**, 89–112.

Bebout, D.G. & Loucks, R.G. (1974) Stuart City trend, Lower Cretaceous, South Texas. *Rep. Bur. Econ. Geol. Univ. Texas Austin* 78, 80 pp.

Bebout, D.G., Budd, D.A. & Schatzinger, R.A. (1981) Depositional and diagenetic history of the Sligo and Hosston formations (Lower Cretaceous) in South Texas. *Rep. Bur. Econ. Geol. Univ. Texas Austin* 109, 70 pp.

Bilotte, M. (1985) Le Crétacé supérieur des plates-formes est-pyrénéennes. *Strata: Actes du Laboratoire de géologie sedimentaire et paléontologie de l'Université Paul-Sabatier, Toulouse*, Série 2, Mémoire 5.

Borgomano, J. & Philip, J. (1989) The rudist carbonate build-ups and the gravitary carbonates of the Gargano-Apulian margin (southern Italy, Upper Senonian). In: *Proceedings of the International Symposium on Evolution of the Karstic Carbonate Platform: Relation with Other Peri-Adriatic Carbonate Platforms. Trieste, 1–6 June 1987* (Eds G.B. Carulli, F. Cucchi & C. Pirini-Radrizzani). Mem. Soc. Geol. Italiana 40, 125–132.

Burchette, T.P. & Britton, S.R. (1985) Carbonate facies analysis in the exploration for hydrocarbons: a case study from the Cretaceous of the Middle East. In: *Sedimentology: Recent Developments and Applied Aspects* (Eds P.J. Brenchley & B.P.J. Williams). Spec. Publ. Geol. Soc. Lond. 18, 311–338.

Camoin, G., Bernet-Rollande, M.-C. & Philip, J. (1988) Rudist–coral frameworks associated with submarine volcanism in the Maastrichtian of the Pachino area (Sicily). *Sedimentology* **35**, 123–138.

Camoin, G., Bouju, J.-P., Maurin, A.-F., Perthuisot, V. & Rouchy, J.-M. (1990) Relations récifs-diapirs: l'exemple du Sénonien de la région de Khenchela (Algérie centre-orientale). *Bull. Soc. Geol. France* Série 8, **6**, 831– 841.

Carbone, F. & Sirna, G. (1981) Upper Cretaceous reef models from Rocca di Cave and adjacent areas in Latium, Central Italy. In: *European Fossil Reef Models* (Ed. D.F. Toomey). Spec. Publ. Soc. Econ. Paleont. Miner. 30, 427–445.

Carulli, G.B., Cucchi, F. & Pirini-Radrizzani, C. (Eds.) (1989) *Proceedings of the International Symposium on Evolution of the Karstic Carbonate Platform: Relation with other Peri-Adriatic Carbonate Platforms. Trieste, 1–6 June 1987.* Mem. Soc. Geol. Italiana 40.

Collins, L.S. (1988) The faunal structure of a mid-Cretaceous rudist reef core. *Lethaia* **21**, 271–280.

Cowen, R. (1983) Algal symbiosis and its recognition in the fossil record. In: *Biotic Interactions in Recent and Fossil Benthic Communities* (Eds M.J.S. Tevesz & P.L. McCall). Topics in Geobiology 3, pp. 431–479. Plenum Publishing: New York.

Drobne, K., Ogorelec, B., Pleničar, M., Zucchi-Stolfa, M.L. & Turnšek, D. (1988) Maastrichtian, Danian and Thanetian beds in Dolenja Vas (NW Dinarides, Yugoslavia). Microfacies, foraminifers, rudists and corals. *Academia scientarum et artium Slovenica, Classis IV: historia naturalis: Dissertationes* 29. Slovenian Academy: Llubljana.

Enos, P. (1977) Tamabra Limestone of the Poza Rica trend, Cretaceous, Mexico. In: *Deep-water Carbonate Environments* (Eds H.E.Cook & P. Enos). Spec. Publ. Soc. Econ. Paleont. Mineral. Tulsa 25, 273–314.

Enos, P. (1988) Evolution of pore-space in the Poza Rica trend (mid-Cretaceous), Mexico. *Sedimentology* **35**, 287–325.

Fagerstrom, J.A. (1987) *The Evolution of Reef Communities.* J. Wiley & Sons: New York.

Fernandez-Mendiola, P.A. & Garcia-Mondejar, J. (1989) Sedimentation of a Lower Cretaceous (Aptian) coral mound complex, Zaraya Mountains, northern Spain. *Geol. Mag.* **126**, 423–434.

Frost, S.H., Bliefnick, D.M. & Harris, P.M. (1983) Deposition and porosity evolution of a Lower Cretaceous rudist buildup, Shuaiba Formation of eastern Arabian Peninsula. In: *Carbonate Buildups – a Core Workshop* (Ed. P.M. Harris). Soc. Econ. Paleont. Mineral. Core Workshop, 4, 381–410.

Gili, E. (1984) *Interaccions sedimentològiques i biològiques a les formacions calcàries de rudistes (Bivalvia) de les Collades de Basturs (Cretaci Superior, Zona Sudpirinenca Central).* Resum de Tesi Doctoral, Universitat Autònoma de Barcelona, Bellaterra.

Gili, E., Masse, J.-P. & Skelton, P.W. (1990). Did rudists build reefs? *Second International Conference on Rudists, October 1990 (Abstracts).* p. 9. Dipartimento di Scienze

della Terra, Università 'La Sapienza': Rome, and Dipartimento di Geologia e Geofisica, Università degli Studi: Bari, Italy.

Grosheny, D. & Philip, J. (1989). Dynamique biosédimentaire de bancs à rudistes dans un environnement péri-deltaïque: la formation de la Cadière d'Azur (Santonien, SE France). *Bull. Soc. Géol. France* Série 8, **5**, 1253–1269.

Gušić, I. & Jelaska, V. (1990) *Upper Cretaceous Stratigraphy of the Island of Brač Within the Geodynamic Evolution of the Adriatic Carbonate Platform.* Jugoslavenska akademija znanosti i umjetnosti Institu za geološka istraživanja, OOUR za geolojiju Zagreb, 69 (in Serbo-Croat with English summary).

Hallock, P. & Schlager, W. (1986) Nutrient excess and the demise of coral reefs and carbonate platforms. *Palaios* **1**, 389–398.

Hancock, J.M. & Kauffman, E.G. (1979) The great transgressions of the late Cretaceous. *J. Geol. Soc. Lond.* **136**, 175–186.

Haq, B.V., Hardenbol, J. & Vail, P.R. (1987) Chronology of fluctuating sea-levels since the Triassic. *Science* **235**, 1156–1167.

Harris, P.M. & Frost, S.H. (1984) Middle Cretaceous carbonate reservoirs, Fahud field and NW Oman. *Bull. Am. Ass. Petrol. Geol.* **68**, 649–658.

Höfling, R. (1985) Faziesverteilung und Fossilvergesellschaftungen im karbonatischen Flachwasser-Milieu der alpinen Oberkreide (Gosau-Formation). *Münchner Geowiss. Abh.,* (A)**3**, 241 pp.

Huffington, T.L. (1981) *Faunal Zonation and Hydrothermal Diagenesis of a Cenomanian (Middle Cretaceous) Rudist Reef, Paso del Rio, Colima, Mexico.* Unpublished MS Thesis, University of Texas: Austin.

Jackson, J.B.C. (1985) Distribution and ecology of clonal and aclonal benthic invertebrates. In: *Population Biology and Evolution of Clonal Organisms* (Eds J.B.C. Jackson, L.W. Buss & R.E. Cook) pp. 297–355. Yale University Press: Connecticut.

James, N.P. (1984) Reefs. In: *Facies Models,* 2nd edn (Ed. R.G. Walker). Geosci. Canada Rep. Ser. 1, 229–244. Geological Association of Canada: Toronto.

Kauffman, E.G. & Johnson, C.C. (1988) The morphological and ecological evolution of Middle and Upper Cretaceous reef-building rudistids. *Palaios* **3**, 194–216.

Konishi, K. (1985) Cretaceous reefal fossils dredged from two seamounts of the Ogasawara Plateau. In: *Preliminary Report of the Hakuhō Maru Cruise KH 84–1* (Ed. K. Kobayashi) pp. 169–179. Ocean Research Institute, University of Tokyo: Tokyo.

Luperto Sinni, E. & Masse, J.-P. (1982) Contributo della paleoecologia alla paleogeografia della parte meridionale della piattaforma Apula nel Cretaceo inferiore. *Geol. Romana* **21**, 859–877.

Masse, J.-P. (1979a) Schizophytoides du Crétacé inférieur. Caractéristiques et signification écologique. *Bull. Cent. Rech. Expl. Prod. Elf-Aquitaine* Pau, France, **3**, 685–703.

Masse, J. -P., (1979b) Les rudistes (Hippuritacea) du Crétacé Inférieur. Approche paléoécologique. In: *L'Urgonien des pays méditerranéens* (Eds A. Arnaud-Vanneau & H. Arnaud). Géobios Mémoire Spécial 3, 277–287.

Masse, J.-P. (1989) Relations entre modifications biologiques et phénomènes géologiques sur les plates-formes carbonatée du domaine périmediterranéen au passage Bédoulien-Gargasien. In: *Les événements de la partie moyenne du Crétacé (Aptien à Turonien)* (Ed. P. Cotillon). Géobios Mémoire Spécial, 11, 279–294.

Masse, J.-P. & Luperto-Sinni, E. (1989) A platform to basin transition model: the Lower Cretaceous carbonates of the Gargano massif (southern Italy). In: *Proceedings of the International Symposium on Evolution of the Karstic Carbonate Platform: Relation with Other Peri-Adriatic Carbonate Platforms. Trieste, 1–6 June 1987* (Eds G.B. Carulli, F. Cucchi & C. Pirini-Radrizzani). Mem. Soc. Geol. Italiana 40, 99–108.

Masse, J.-P. & Philip, J. (1981) Cretaceous coral–rudistid buildups of France. In: *European Fossil Reef Models* (Ed. D.F. Toomey). Spec. Publ. Soc. Econ. Paleont. Miner. Tulsa, 30, 399–426.

Masse, J.-P. & Philip, J. (1986) L'évolution des rudistes au regard des principaux événements géologiques du Crétacé. *Bull. Cent. Rech. Expl. Prod. Elf-Aquitaine* 10, 437–456.

Mennessier, G. (1949) Sur la présence de Rudistes dans un sédiment hautement ligniteux à Piolenc (Vaucluse). *Comp.-rend. Somm. Séan. Soc. Géol. France* 14, 215–217.

Philip, J., Airaud, C. & Tronchetti, G. (1989) Événements paléogéographiques en Provence (SE France) au passage Cénomanien–Turonien, modifications sédimentaires-causes géodynamiques. In: *Les événements de la partie moyenne du Crétacé (Aptien à Turonien)* (Ed. P. Cotillon).Géobios Mémoire Spécial, 11, 107–117.

Pieri, P. & Laviano, A. (1989) Tettonica e sedimentazione nei depositi senoniani delle Murge sud-orientali (Ostuni). *Boll. Soc. Geol. Italiana* 108, 351–356.

Ross, D.J. (1992) Sedimentology and depositional profile of a mid-Cretaceous shelf edge rudist reef complex, Nahal Ha'mearot, NW Israel. *Sedim. Geol.* 78, 156–168.

Schlager, W. (1981) The paradox of drowned reefs and carbonate platforms. *Bull. Geol. Soc. Am.* 92, 197–211.

Scholle, P.A. & Arthur, M.A. (1980) Carbon isotope fluctuations in Cretaceous pelagic limestones: potential stratigraphic and petroleum exploration tool. *Bull. Am. Ass. Petrol. Geol.* 64, 67–87.

Scott, R.W. (1988) Evolution of Late Jurassic and Early Cretaceous reef biotas. *Palaois* 3, 184–193.

Scott, R.W. (1990) Models and stratigraphy of mid-Cretaceous reef communities, Gulf of Mexico. *Soc. Econ. Paleon. Miner. Conc. Sedim. Paleont.* 2, 102 pp.

Scott, R.W., Fernandez-Mendiola, P.A., Gili, E. & Simo, A. (1990) Persistance of coral–rudist reefs into the Late Cretaceous. *Palaios* 5, 98–110.

Simmons, M.D. & Hart, M.B. (1987) The biostratigraphy and microfacies of the early to mid-Cretaceous carbonates of Wadi Mi'aidin, Central Oman Mountains. In: *Micropalaeontology of Carbonate Environments* (Ed. M.B. Hart) pp. 176–207. Ellis Horwood Ltd: Chichester.

Sirna, M. & Cestari, R. (1989) Il Senoniano a rudiste (Hippuritacea) del settore sudoccidentale della piattaforma carbonatica Laziale-Abruzzese (Appennino Centrale). *Boll. Soc. Geol. Italian* 108, 711–719.

Skelton, P.W. (1976) Functional morphology of the Hippuritidae. *Lethaia* 9, 83–100.

Skelton, P.W. (1978) The evolution of functional design in rudists (Hippuritacea) and its taxonomic implications. In: *Evolutionary Systematics of Bivalve Molluscs* (Eds M. Yonge & T.E. Thompson). Phil. Trans. R. Soc. Lond. B 284, 305–318.

Skelton, P.W. (1985) Preadaptation and evolutionary innovation in rudist bivalves. In: *Evolutionary Case Histories from the Fossil Record* (Eds J.C.W. Cope & P.W. Skelton). Spec. Pap. Palaeont. 33, 159–173.

Skelton, P.W. (1991) Morphogenetic versus environmental cues for adaptive radiations. In: *Constructional Morphology and Evolution* (Eds N. Schmidt- Kittler & K. Vogel) pp. 375–388. Springer-Verlag: Berlin.

Skelton, P.W. & Gili, E. (1991) Palaeoecological classification of rudist morphotypes. In: *First International Conference on Rudists, October 1988, Proceedings* (Ed. M. Sladić-Trifunović) pp. 71–86. Serbian Geological Society: Belgrade.

Skelton, P.W. & Wright, V.P. (1987) A Caribbean rudist bivalve in Oman: island-hopping across the Pacific in the late Cretaceous. *Palaeontology* 30, 505–529.

Skelton, P.W., Nolan, S.C. & Scott, R.W. (1990) The Maastrichtian transgression onto the northwestern flank of the Proto-Oman Mountains: sequences of rudist-bearing beach to open shelf facies. In: *The Geology and Tectonics of the Oman Region* (Eds A.H.F. Robertson, M.P. Searle & A.C. Ries). Geol. Soc. Lond. Spec. Publ. 49, 521–547.

Swinburne, N.H.M. (1990) *The extinction of the Rudist Bivalves.* Unpublished PhD Thesis, Open University: Milton Keynes.

6   Oxygen-related mudrock biofacies

DAVID J. BOTTJER AND CHARLES E. SAVRDA

Introduction

Although fine-grained strata constitute about 60% of the stratigraphic record (Potter *et al.*, 1980), they have only been studied intensively during the past 20 years. This relatively recent increase in the study and understanding of fine-grained sedimentary rocks has been due to: new microstratigraphic methods (e.g. Savrda & Bottjer, 1989a); detailed SEM analysis of mudrock components (e.g. O'Brien & Slatt, 1990; Bennett *et al.*, 1991); and innovative geochemical approaches (e.g. Raiswell & Berner, 1985; Wignall & Myers, 1988). In contrast, fossils from many fine-grained sedimentary units are unusually well-preserved and have a long history of biostratigraphic and phylogenetic study. A greater understanding of the sedimentology of these lithified muds has led to a heightened scrutiny of the palaeoecology of mudrock faunas, particularly those deposited in oxygen-deficient environments.

The term mudrock is used broadly in this review so as to include sedimentary rocks that originated as fine-grained sediments, with an original composition dominated by clay, carbonate or silica. Similarly, biofacies are defined as distinctive suites of biological characteristics of strata that are indicative of specific palaeoenvironmental conditions (e.g. degree of benthic oxygenation), under which those particular rocks were deposited.

Much of the biotic and sedimentary history of Earth's marine environments is recorded in mudrock biofacies. In particular, the study of oxygen-related mudrock biofacies has proved to be indispensable in deciphering the physical and biological dynamics of many ancient epicontinental seas, which were much more widespread in the past than they are now. Because most hydrocarbon source rocks were deposited in oxygen-deficient palaeoenvironments, an understanding of oxygen-related mudrock biofacies is also crucial in the search for and production of petroleum.

Early studies on functional morphology of mudrock faunas

Marine mudrock faunas commonly contain fossils with unusual morphologies indicative of a stressed environment. Early work by Rhoads & Young (1970) on modern muddy-bay environments demonstrated the principle of trophic group amensalism, whereby the presence of suspension feeders is limited by the activity of abundant deposit feeders. Such limiting activity by deposit feeders includes bioturbation and subsequent fluidization of sea-floor surfaces, so that suspension feeders cannot attain a stable living surface (for an extended discussion on the effects of this process, see Thayer, 1983). This principle has been widely applied to mudrock faunal assemblages in the fossil record (e.g. Thayer, 1983) to explain the rarity or even lack of suspension-feeding components.

Rhoads (1970, 1974) and others (e.g. Carter, 1968; Stanley, 1970; Surlyk, 1972; Thayer, 1975) were also working on a functional understanding of morphological adaptations that epifaunal suspension feeders use to live on muddy substrates. From this work arose an understanding of several morphological strategies for life on soft, muddy sea-floors. Such strategies include: (i) small adult size because high surface-to-volume ratios and thin shells enhance floatation on soft substrates, (ii) colonization of relatively large, hard substrates projecting above the fluid muddy substrate, (iii) the snowshoe strategy, which involves development of adult morphologies with a broad, flat form and/or radiating skeletal projections that distribute the weight of the organism over a large surface area, and (iv) the iceberg strategy, which involves development of adult morphologies with an expanded or inflated ventral surface, resulting in the classic gryphaete shape, whereby the ventral part of the organism projects through the more fluid surface layer to firmer underlying sediment. This functional morphologic conceptual framework has been extensively applied to fossil faunas, particularly those of Upper Cretaceous chalks, which were originally deposited on soft, carbonate mud sea-floors (e.g. Carter, 1972; Surlyk, 1972; Bottjer, 1981; Jablonski & Bottjer, 1983).

Early studies on oxygen-related mudrock biofacies

Black shales comprise that subset of mudrocks that are laminated and/or fissile. Geologists and palaeontologists have wrestled for a long time to understand the oxygen-deficient environments that lead to the deposition of black shales and the sometimes remarkably well-preserved fossil faunas that are found within them (e.g. see summary by Dunbar & Rogers, 1957, of early literature). These fossil faunas typically exhibit a mixture of planktonic, pseudoplanktonic, nektonic, and benthic life habits. Traditional interpretations of such faunas understood the stratified nature of many modern oxygen-deficient basins, and utilized a general principle that benthic animals should not be able to live on sea-floors where black shales are being deposited. These interpretations thus tended to reconstruct the life habits of all fossils found in black shales to be planktonic, pseudoplanktonic, or nektonic, including fossils that, if they were found in other sedimentary rock types, would be interpreted as benthic. Such 'typically' benthic fossils were traditionally reassigned to a pseudoplanktonic or sometimes a nektonic life habit (e.g. Jefferies & Minton, 1965).

Rhoads & Morse (1971) made an excellent synthesis of data on extant oxygen-deficient basins, in order to better understand the role that increasing oxygen concentrations (which they reported as ml/l at STP) may have had in the early Phanerozoic history of the Metazoa. This synthesis was utilized, in a landmark paper on black shales by Byers (1977), to develop criteria for recognizing three oxygen-related biofacies in the stratigraphic record. In the Rhoads–Morse–Byers model, marine environments with greater than 1.0 ml/l (STP) of dissolved oxygen typically produce a sedimentary record characterized by abundant bioturbation and calcareous body fossils: these conditions result in deposition of the aerobic biofacies. A somewhat oxygen-deficient sea-floor environment, with oxygen concentrations between 1.0 ml/l and 0.1 ml/l (STP), is interpreted in the Rhoads–Morse–Byers model to lead to deposition of the dysaerobic biofacies, which they described as characterized by

a partially bioturbate sedimentary fabric with poorly calcified benthic faunas dominated by deposit feeders. The concept of the dysaerobic biofacies has received wide acceptance in the study of oxygen-deficient basins (e.g. Kammer et al., 1986).

The biofacies that represents the lowest oxygen concentrations, the anaerobic biofacies (oxygen concentrations lower than 0.1 ml/l (STP)), is defined in the Rhoads–Morse–Byers model as undisturbed (laminated) sediment lacking all benthos. This definition for an anaerobic biofacies tended to reinforce older ideas that 'typically' benthic fossils found in laminated black shale strata could not be *in situ*, but must have been transported to their final place of deposition from an overlying better-oxygenated water mass or by processes such as turbidity currents or debris flows. Further detailed investigations into the exact nature of what the Rhoads–Morse–Byers model would term the anaerobic biofacies have served to drive much of the recent work on mudrock biofacies.

The Jurassic Posidonienschiefer: a classic puzzle for oxygen-deficient mudrock biofacies

The Toarcian Posidonienschiefer of southern Germany has been studied extensively, in large part due to its exquisitely-preserved remains of Mesozoic marine reptiles (e.g. Hauff & Hauff, 1981). This fossil-lagerstätte was deposited in an epicontinental sea setting and also contains abundant fossils that are typically assigned to a benthic life habit (particularly the bivalves '*Posidonia*' (= *Bositra*) *radiata* and *Pseudomytiloides dubius*). Because this unit is generally a laminated black shale, however, all 'typically' benthic fossils had been interpreted by earlier workers to be either nektonic or pseudoplanktonic (see Kauffman, 1981, for a summary of this earlier work).

Perhaps the best-known example is that of '*Posidonia*'. Based on its facies distribution, primarily in Jurassic black shales, as well as considerations of functional morphology, Jefferies & Minton (1965) considered *Bositra buchi*, a close relative of '*Posidonia*', to be nektoplanktonic. This analysis was accompanied by a comparative study of '*Posidonia*' *radiata*, which Jefferies & Minton (1965) also suggested had been nektoplanktonic. Kauffman (1981), however, contested their functional morphologic interpretations, and concluded that these taxa were truly benthic.

The Posidonienschiefer contains spectacular assemblages of fossils associated with fossil logs. These logs are commonly encrusted with the inoceramid bivalve *Pseudomytiloides dubius* and also have attached complete specimens of the large pentacrinitid crinoid *Seirocrinus subangularis*. In earlier studies much debate centered on the timing of colonization of these fossil logs by invertebrates, with much of the evidence also coming from studies of functional morphology. One school held that the logs were colonized as driftwood in an overlying better-oxygenated part of the water column (e.g. Seilacher *et al.*, 1968), while the other school postulated that the logs were only colonized by invertebrates after settling on the sea-floor (e.g. Rasmussen, 1977; Kauffman, 1981).

This controversy over life habits of Posidonienschiefer fossil invertebrates characterizes the palaeoecological puzzle of 'typically' benthic macroinvertebrates found in laminated black shales. Depending upon these interpretations of life habits for Posidonienschiefer fossils, this unit was either deposited wholly within an anaerobic biofacies, a dysaerobic biofacies, or some combination of both. Distribution in black shales and data from studies of functional morphology have commonly been cited as the reasons for assigning these fossils a nektonic or pseudoplanktonic life habit. It is apparent, as summarized above, that functional morphology commonly lacks the resolving power to determine life habit. Thus facies distribution, potentially also a weak link in a life-habit argument, was the primary line of evidence. Clearly, additional evidence from other sources was needed.

New lines of evidence: Miocene Monterey Formation

Coincident with a study that we were completing on Posidonienschiefer trace fossils (Savrda & Bottjer, 1989b), we had also been examining the Miocene Monterey Formation of California, much of which is interpreted to have been deposited as siliceous mud in oxygen-deficient environments (e.g. Pisciotto & Garrison, 1981). Our studies of the Monterey were primarily focused on developing a trace fossil model for reconstructing oxygen-deficient palaeoenvironments (Savrda & Bottjer, 1986, 1987a, 1989a). Several stratigraphic sections were under examination, but, in particular we studied the section along Toro Road (late Miocene Canyon del Rey Member) in Monterey County,

California. This section was chosen because, during an earlier visit to this outcrop, R. Garrison & J. Ingle (pers. comm., 1983) had demonstrated that large specimens of the bivalve *Anadara montereyana* exist there in diatomaceous deposits, a Monterey facies that commonly is interpreted to have been deposited under conditions with extremely low levels of dissolved oxygen. However, given our then understanding of oxygen-deficient biofacies, derived from the Rhoads–Morse–Byers model, the occurrence of *Anadara montereyana* from the Toro Road section thus seemed to indicate that an aerobic biofacies was present. We focused on this deposit because we had never studied a Monterey diatomite that had potentially been deposited under aerobic conditions.

Application of a trace fossil model for determining relative amounts of depositional palaeo-oxygenation (see Savrda & Bottjer (1986, 1987a, 1989a) for an in-depth discussion of this model) to a 1-m thick *Anadara montereyana*-bearing interval, however, revealed that this section had in reality been deposited under oxygen-deficient conditions (Savrda & Bottjer, 1987b) (Fig. 6.1). Furthermore, *Anadara montereyana* occurred only in bedding-plane accumulations (Fig. 6.2) at the interfaces between laminated and bioturbated strata (Savrda & Bottjer, 1987b) (Fig. 6.1). From an interpreted palaeo-oxygenation curve, made from independent sediment fabric and trace fossil evidence, we concluded that these bivalves had lived in benthic environments at the dysaerobic/anaerobic boundary, according to the Rhoads–Morse–Byers model (Savrda & Bottjer, 1987b).

The exaerobic biofacies

Thus, 'typical' benthic macroinvertebrates found in black shales can indeed have had a benthic life habit. Using the Rhoads–Morse–Byers model, we determined that this association of benthic macroinvertebrates occurring in bedding-plane accumulations at the dysaerobic/anaerobic biofacies boundary was an oxygen-deficient biofacies that had great significance but which had not formally been defined. We proposed the term 'exaerobic' biofacies for this association, and extended it to also include benthic macroinvertebrates found *within* laminated strata (Savrda & Bottjer, 1987b).

The question remains as to why benthic macroinvertebrates would live in such a presumably hostile habitat. Oxygen levels would need to be

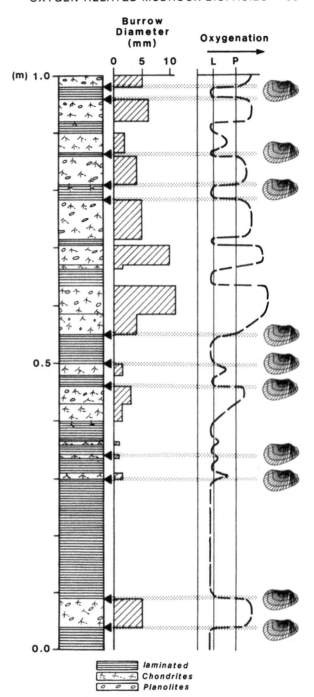

Fig. 6.1 Presentation of data from high-resolution vertical sequence analysis of section of the Monterey Formation located along Toro Road (locality described in Savrda & Bottjer, 1987b). General sedimentary rock fabric types and trace fossil assemblage composition, illustrated schematically in the column, have been used in conjunction with burrow size data to construct the interpreted relative oxygenation curve using the model described in Savrda & Bottjer (1986, 1987a, 1989a). Arrows, stippled bars and schematic *Anadara montereyana* indicate locations of horizons characterized by dense accumulations of large specimens of this bivalve, all of which occur at transitions between anaerobic and more oxygenated strata. (Modified from Savrda & Bottjer, 1987b.)

sufficient for respiration by these metazoans. Indeed, given the nature of oxygenation gradients from the sea-floor to the overlying water column, it is possible that oxygen levels in water directly overlying the sea-floor could have had dysaerobic concentrations. However, such periods of higher oxygen concentrations would probably have been brief, because if they had persisted for any length of

Fig. 6.2 Accumulation of *Anadara montereyana* on bedding plane from section described in Fig. 6.1. Scale in centimetres.

time an infauna would have been expected to colonize the sea-floor and to leave evidence of bioturbation. Because, by definition, evidence for bioturbation does not exist, oxygen levels must have been more typically at the lower end of dysaerobic concentrations.

Such low levels of oxygenation might provide a refuge for benthic macroinvertebrates from predators which require higher levels of oxygenation (Savrda & Bottjer, 1987b). The relatively high levels of organic material found in laminated sediments deposited in oxygen-deficient environments may have also provided a powerful attractant. We have suggested the possibility that macroinvertebrates found in the exaerobic biofacies, like those at modern hydrothermal vents, cold seeps and sewage outfalls, had chemosymbiotic sulphur-oxidizing bacteria (Savrda & Bottjer, 1987b). Chemosymbiosis would enable these organisms to utilize energy from forms of sedimentary organic material that are typically undigestible by macroinvertebrates. Indeed, it seems that a chemosymbiotic modern analogue for the Monterey fossil occurrences exists in the oxygen-deficient Santa Barbara Basin off of southern California, where *Lucinoma aequizonata*, a chemosymbiotic lucinid bivalve, lives in a narrow band around the basin at depths which mark the change in sea-floor dissolved oxygen concentrations from dysaerobic to anaerobic values (Cary *et al.*, 1989). However, until more work is done on living as well as fossil chemosymbiotic macroinvertebrates and the environments which they inhabit, it seems entirely possible that other currently unknown adaptations may allow organisms to live in settings that would produce an exaerobic biofacies in the stratigraphic record.

Exaerobic biofacies in epicontinental seas

The Monterey Formation was deposited in an active continental margin setting, with narrow continental shelves and a mid-water oxygen-minimum zone that impinged upon the slope (Pisciotto & Garrison, 1981). At Toro Road the exaerobic biofacies is restricted to very narrow (less than 1 cm) stratigraphic intervals (Fig. 6.1). This suggests that conditions under which the exaerobic biofacies was deposited were narrowly restricted. As these conditions shifted their position dynamically, in response to fluctuations in the vertical position and/or intensity of the oceanic oxygen-minimum zone, a vertical stratigraphy was produced with commonly narrow alternations between laminated strata, bioturbated strata, and strata with the exaerobic biofacies.

The exaerobic biofacies concept can also be applied to ancient epicontinental seaways. Consider broad, shallow, relatively flat-bottomed seaways characterized by warm bottom waters and high influx of marine and/or terrestrial organic matter. In these settings, oxygen deficiency was more likely generated at the sea-floor rather than in the mid-water column in response to intense bacterial oxygen consumption in the sediment. During times of diminished circulation, the dysaerobic/anaerobic boundary would have risen above the sediment–water interface. During periods of increased bottom-water circulation, it is theoretically possible that a diffusive boundary layer developed, resulting in the stabilization of the dysaerobic/anaerobic boundary directly at the sediment–water interface (Jorgensen & Revsbech, 1985). The development of bacterial mats on the sea-floor also has been proposed as a potential contributing factor towards perching the redox potential discontinuity at the sediment–water interface. These combined factors might have interacted in such a way that bottom-water oxygen levels were sufficient to periodically support epifauna, but redox conditions in the substrate precluded the establishment of bioturbating infauna (Kauffman, 1981; Savrda & Bottjer, 1987b; Sageman, 1989; Savrda et al., 1991). Temporal fluctuations in circulation could then provide a much more laterally extensive and stratigraphically thicker record of exaerobic biofacies deposition than that found in the Monterey. The record of this biofacies in epicontinental seas would be reflected by sequences of laminated, organic-rich strata containing epibenthic macroinvertebrate fossils.

Posidonienschiefer: a re-evaluation

The exaerobic biofacies concept has great potential for resolving some of the conflicting interpretations of Posidonienschiefer benthic macroinvertebrate fossil life habits. Intervals up to several metres thick in the Posidonienschiefer have a laminated sediment fabric and contain abundant accumulations on bedding planes of the bivalve 'Posidonia' radiata (Fig. 6.3). Kauffman's (1981) reanalysis of functional morphology has convincingly argued that 'Posidonia' radiata was an epibenthic bivalve. These intervals thus represent an exaerobic biofacies and 'Posidonia' radiata may have had a bacterial chemosymbiosis (Kauffman, 1988; Savrda et al., 1991).

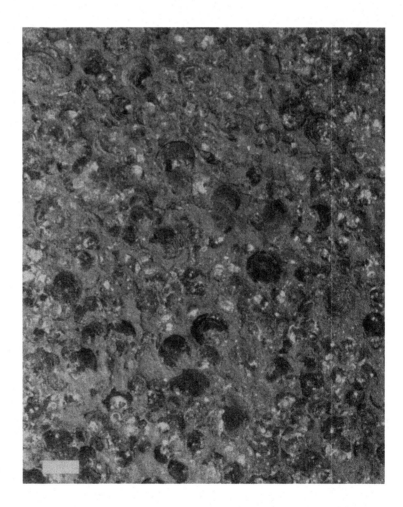

Fig. 6.3 Bedding-plane accumulation of 'Posidonia' radiata from the Posidonienschiefer at Holzmaden, Germany. Scale bar is 10 mm.

The exaerobic biofacies concept is not equally applicable to all occurrences of 'typical' benthic macroinvertebrates found within laminated strata. This is illustrated by the case of *Seirocrinus subangularis* found attached to logs in the Posidonienschiefer. Extensive analysis of many pentacrinitids by Simms (1986) shows that in fact this was a pseudoplanktonic group of crinoids, adapted to living on driftwood (see Wignall & Simms, 1990, for an in-depth discussion of pseudoplankton in the fossil record). Its occurrence in the Posidonienschiefer is therefore not evidence for an exaerobic biofacies.

The inoceramid *Pseudomytiloides dubius* thus remains as the last part of the contested puzzle on 'typical' Posidonienschiefer benthic macroinvertebrate fossil life habits. *Pseudomytiloides dubius*, in association with *Seirocrinus subangularis*, is certainly found attached to logs in this unit. If the logs were driftwood when they were colonized by *Seirocrinus subangularis*, there is no reason to believe that *Pseudomytiloides dubius* also did not colonize them when they were driftwood. This establishes the likelihood that at least some *Pseudomytiloides dubius* were pseudoplanktonic. Kauffman (1981), however, has argued that *Pseudomytiloides dubius* is too abundant within the Posidonienschiefer to have originated from drifting logs. This point is sufficiently convincing since fossil logs, in fact, are rare. Hence, the possibility that some, if not many, *Pseudomytiloides dubius* were epibenthic needs further consideration (Wignall & Simms, 1990).

Other Jurassic black shales

Morris (1979), in a synthesis of Jurassic marine shale sequences from the Toarcian of Great Britain, described three types of shale, including a bituminous shale that is finely laminated, with little or no bioturbation and a sparse epibenthic fauna. Oschmann (1988) and Wignall & Myers (1988), in paleoecological studies of the Upper Kimmeridge Clay of southern England, also described a bituminous shale with microlamination that lacks bioturbation but contains infaunal and epibenthic macroinvertebrates. These and other bituminous shale intervals in both the Toarcian and Kimmeridgian deserve consideration as representing the exaerobic biofacies.

Devonian Geneseo Shale

The Upper Devonian Geneseo Shale from central New York represents a classic epicontinental sea, black shale facies. In early studies, the life habits of the brachiopod-dominated fauna of 'typically' benthic macroinvertebrate fossils were interpreted to be pseudoplanktonic. For example, Thayer (1974), in an elegant study of central New York Upper Devonian palaeoecology, interpreted Geneseo faunas dominated by the brachiopod *Leiorhyncus* to have been pseudoplanktonic upon seaweed before deposition in black shale environments. However, a recent detailed study of this fauna by Thompson & Newton (1987) indicates that the *Leiorhyncus* fauna, found within laminated strata, is indeed epibenthic. Thompson & Newton (1987) describe a depositional environment for this fauna that is almost identical to that described by Savrda & Bottjer (1987b) for the exaerobic biofacies in the Monterey Formation; the exaerobic biofacies thus appears to exist in the Geneseo Shale.

Cretaceous Smoky Hill Chalk

The Upper Cretaceous Smoky Hill Chalk member of the Niobrara Formation, which was deposited in an oxygen-deficient epicontinental seaway that covered the western interior of North America, has been studied most intensively in western Kansas and Colorado (e.g. Hattin, 1982). Throughout much of the Smoky Hill occur large (up to 2.0 m), flat relatively isolated specimens of the bivalve *Inoceramus platinus*, which exhibit a snowshoe adaptation to living on soft substrates. These inoceramids are typically found in laminated strata, so that their occurrence probably indicates the exaerobic biofacies (Savrda & Bottjer, 1987b). These relatively isolated occurrences within the Smoky Hill differ from occurrences of other exaerobic faunas, which are common as shell pavements along bedding planes (e.g. '*Posidonia*' *radiata* in the Posidonienschiefer; Fig. 6.3).

Further subdivision of the original anaerobic biofacies

Since the development of the exaerobic biofacies concept it has become apparent that the original Rhoads–Morse–Byers anaerobic biofacies, which included all fine-grained strata characterized by laminations, can be subdivided further. As originally defined, bottom waters that lead to deposition of the anaerobic biofacies are not necessarily anoxic, but may be characterized by extremely low concentrations of dissolved oxygen (between 0.0 and 0.1 ml/l). Despite the exclusion of bioturbating macrobenthic organisms at these low levels of oxygenation, these environments may host preservable benthic microfauna, such as Foraminifera, which have been documented from laminated sediments of several modern oxygen-deficient settings (e.g. Douglas, 1979). Ancient analogues of these assemblages have also been recognized from laminated strata of a variety of ages (e.g. Govean & Garrison, 1981; Koutsoukos *et al.*, 1990). In addition to benthic Foraminifera, environments with sea-floor dissolved oxygen concentrations between 0.0 and 0.1 ml/l may host other very small soft-bodied organisms, such as nematodes and sedentary polychaetes (e.g. Hartman & Bernard, 1958; Calvert, 1964). Although these meiofaunal organisms are generally characterized by low preservation potential, scolecodonts of benthic polychaetes have recently been identified in various Mesozoic laminated strata (Courtinat & Howlett, 1990). Activities of these organisms may result in a subtle, incomplete disruption of primary lamination: a sedimentary fabric that has been termed microbioturbation (Pratt, 1984). Microbioturbation has been found in the anaerobic zones of modern California borderland basins (Savrda *et al.*, 1984), as well as in ancient laminated strata (e.g. Pratt, 1984; Sageman, 1989).

Truly anoxic sea-floor environments (bottom water dissolved oxygen of zero) are characterized by laminated sediments but lack *in situ* benthic Foraminifera and active organisms capable of producing microbioturbation. Hence, two distinct biofacies can be discerned in the stratigraphic record: (i) laminated strata characterized by microbioturbation and/or benthic microfossils, and (ii) laminated strata lacking both of these features. Because the anaerobic biofacies has traditionally been used to indicate deposition under the lowest levels of dissolved oxygen, this term is retained for the biofacies characterized only by laminated strata (Savrda & Bottjer, 1991). The designation 'quasi-anaerobic biofacies' was first used by Koutsoukos *et al.* (1990) for strata deposited in environments with dissolved oxygen concentrations between 0.0 and 0.1 ml/l. The quasi-anaerobic biofacies is appropriate for strata where anoxic conditions are apparent on the basis of the presence of lamination and the absence of *in situ* macrobenthic body fossils, but where closer scrutiny reveals a microfauna and/or microbioturbation which indicate that conditions were not completely anoxic.

Thus, based on the presence/absence of microbioturbation as well as *in situ* benthic microfossils and macrobenthic body fossils, three biofacies, exaerobic, quasi-anaerobic, and anaerobic, can be recognized from laminated black shales. Characteristics of these three biofacies, as well as the dysaerobic and aerobic biofacies, are summarized in Fig. 6.4.

Discussion

The fascinating nature of oxygen-deficient mudrock biofacies has attracted much interest from palaeoecologists and sedimentologists, and this has led to further postulation of additional biofacies. For example, Sageman (1989) describes a gradient of dissolved oxygen and substrate consistency in a 'benthic boundary biofacies model' to interpret ancient oxygenation conditions of the Upper Cretaceous Hartland Shale Member of the Greenhorn Formation, which was deposited in the North American Western Interior Seaway. Finer subdivision of oxygen-related mudrock biofacies will no doubt occur as more is learned about the detailed properties of mudrock biotas.

A particularly promising avenue of research is the examination of ecological and evolutionary changes in the faunas that have inhabited oxygen-related mudrock biofacies. One long-term trend that is apparent is that epifaunal suspension feeders, particularly bivalves with snowshoe and iceberg adaptations, became much less common with the onset of the Cenozoic (e.g. Jablonski & Bottjer, 1983; Kauffman, 1988; Wignall, 1990). Some of these epifaunal mudrock bivalves, such as the inoceramids, are now extinct, so that relatively little is known about their palaeoecology. Palaeoecology of such taxa unknown from modern environments presents special challenges. For example, it has been suggested that some inoceramids found in black shales may have been chemosymbiotic (Savrda & Bottjer, 1987b; Kauffman, 1988), but confirmation of this hypothesis awaits further investigation, perhaps by isotopic studies of inoceramid shell organic matrices (e.g. Cobabe, 1990). Similarly, it has been proposed that many types of the common black shale trace fossil *Chondrites* were constructed by soft-bodied chemosymbiotic organisms (Savrda *et al.*, 1991). Tests of these and other hypotheses about life in oxygen-deficient palaeoenvironments will have a profound impact upon future studies of oxygen-related mudrock biofacies in the stratigraphic record.

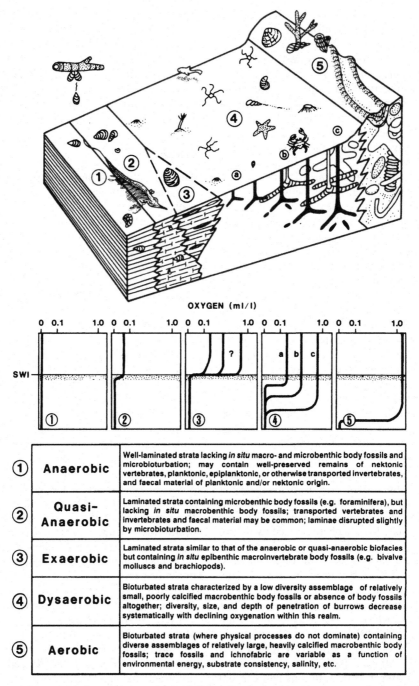

OXYGEN (ml/l)

①	**Anaerobic**	Well-laminated strata lacking *in situ* macro- and microbenthic body fossils and microbioturbation; may contain well-preserved remains of nektonic vertebrates, planktonic, epiplanktonic, or otherwise transported invertebrates, and faecal material of planktonic and/or nektonic origin.
②	**Quasi-Anaerobic**	Laminated strata containing microbenthic body fossils (e.g. foraminifera), but lacking *in situ* macrobenthic body fossils; transported vertebrates and invertebrates and faecal material may be common; laminae disrupted slightly by microbioturbation.
③	**Exaerobic**	Laminated strata similar to that of the anaerobic or quasi-anaerobic biofacies but containing *in situ* epibenthic macroinvertebrate body fossils (e.g. bivalve molluscs and brachiopods).
④	**Dysaerobic**	Bioturbated strata characterized by a low diversity assemblage of relatively small, poorly calcified macrobenthic body fossils or absence of body fossils altogether; diversity, size, and depth of penetration of burrows decrease systematically with declining oxygenation within this realm.
⑤	**Aerobic**	Bioturbated strata (where physical processes do not dominate) containing diverse assemblages of relatively large, heavily calcified macrobenthic body fossils; trace fossils and ichnofabric are variable as a function of environmental energy, substrate consistency, salinity, etc.

Fig. 6.4 Summary of characteristics of oxygen-related biofacies, illustrating lateral oxygen gradients along the sea-floor (top) and vertical oxygen-gradients across the sediment–water interface (SWI) (middle). Schematic oxygen profiles (at STP) are intended to reflect the relationships between bottom-water oxygenation and the time-averaged position of the redox potential discontinuity (the boundary below which dissolved oxygen does not exist). (See Savrda & Bottjer, 1986, 1987a,b, 1989a,b, 1991.) (From Savrda & Bottjer, 1991.)

Acknowledgements

Our research on oxygen-related biofacies was supported by NSF grant EAR-8508970. We thank D.S. Gorsline, A.G. Fischer, A. Seilacher, B. Sageman, P.B. Wignall and K.A. Campbell for discussion during development of the concepts presented in this chapter. P.B. Wignall and P. Allison provided helpful comments on an earlier version of this chapter.

References

Bennett, R.H., Bryant, W.R. & Hulbert, M.H. (1991) *Microstructure of Fine-Grained Sediments*. Springer-Verlag: New York.

Bottjer, D.J. (1981) Structure of Upper Cretaceous chalk benthic communities, southwestern Arkansas. *Palaeogeogr. Palaeoclim. Palaeoecol.* **34**, 225–256.

Byers, C.W. (1977) Biofacies patterns in euxinic basins: a general model. In: *Deep-Water Carbonate Environments* (Eds H.E. Cook & P. Enos). Soc. Econ. Paleont. Miner. Spec. Publ. 25, 5–17.

Calvert, S.E. (1964) Factors affecting distribution of laminated diatomaceous sediments in the Gulf of California. In: *Marine Geology of the Gulf of California* (Eds T.H. Vandel & G.C. Shor). Am. Ass. Petrol. Geol. Mem. 3, 311–330.

Carter, R.M. (1968) Functional studies on the Cretaceous oyster *Arctostrea*. *Palaeontology* **11**, 458–485.

Carter, R.M. (1972) Adaptations of British Chalk Bivalvia. *J. Paleont.* **46**, 325–340.

Cary, S.C., Vetter, R.D. & Felbeck, H. (1989) Habitat characterization and nutritional strategies of the endosymbiont-bearing bivalve *Lucinoma aequizonata*. *Mar. Ecol. Prog. Ser.* **55**, 31–45.

Cobabe, E.A. (1990) Detection of chemosymbiosis in the fossil record: the use of stable isotopes in the organic matrix of lucinid bivalves. *Geol. Soc. Am. Abstr. Progr.* **22**, A36.

Courtinat, B. & Howlett, P. (1990) Dorvilleids and arabellids (Annelida) as indicators of dysaerobic events in well-laminated non-bioturbated deposits of the French Mesozoic. *Palaeogeogr. Palaeoclim. Palaeoecol.* **80**, 145–151.

Douglas, R.G. (1979) Benthic foraminiferal ecology and paleoecology: a review. In: *Foraminiferal Ecology and Paleoecology* (Eds J.H. Lipps, W.H. Berger, M.A. Buzas, R.G. Douglas & C.A. Ross). Soc. Econ. Paleont. Miner. Short Course Notes 6, 21–53.

Dunbar, C.O. & Rogers, J. (1957) *Principles of Stratigraphy*. Wiley: New York.

Govean, F.M. & Garrison, R.E. (1981) Significance of laminated and massive diatomites in the upper part of the Monterey Formation, California. In: *The Monterey Formation and Related Siliceous Rocks of California* (Eds R.E. Garrison, R.G. Douglas, K.E. Pisciotto & C.M. Isaacs). Soc. Econ. Paleont. Miner. Pac. Sect. Spec. Publ. 15, 181–198.

Hartman, O. & Bernard, J.V. (1958) *The Benthic Fauna of the Deep Basins off Southern California*. Allan Hancock Pacific Expeditions 22.

Hattin, D.E. (1982) Stratigraphy and depositional environment of Smoky Hill Chalk Member, Niobrara Chalk (Upper Cretaceous) of western Kansas. *Kansas Geol. Surv. Bull.* **225**, 108 pp.

Hauff, B. & Hauff, R.B. (1981) *Das Holzmadenbuch*. Repro-Druck: Felbach, Germany.

Jablonski, D. & Bottjer, D.J. (1983) Soft-bottom epifaunal suspension-feeding assemblages in the late Cretaceous: implications for the evolution of benthic paleocommunities. In: *Biotic Interactions in Recent and Fossil Benthic Communities* (Eds M.J.S. Tevesz & P.L. McCall) pp. 747–812. Plenum Press: New York.

Jefferies, R.P.S. & Minton, P. (1965) The mode of life of two Jurassic species of '*Posidonia*' (Bivalvia). *Palaeontology* **8**, 156–185.

Jorgensen, T.W. & Revsbech, N.P. (1985) Diffusive boundary layers and the oxygen uptake of sediments and detritus. *Limnol. Oceanogr.* **30**, 111–122.

Kammer, T.W., Brett, C.E., Boardman, D.R. & Mapes, R.H. (1986) Ecologic stability of the dysaerobic biofacies during the late Paleozoic. *Lethaia* **19**, 109–121.

Kauffman, E.G. (1981) Ecological reappraisal of the German Posidonienschiefer (Toarcian) and the stagnant basin model. In: *Communities of the Past* (Eds J. Gray, A.J. Boucot & W.B.N. Berry) pp. 311–381. Hutchinson Ross: Pennsylvania.

Kauffman, E.G. (1988) The case of the missing community: low-oxygen adapted Paleozoic and Mesozoic bivalves ('flat clams') and bacterial symbiosis in typical Phanerozoic seas. *Geol. Soc. Am. Abstr. Progr.* **20**, A48.

Koutsoukos, E.A.M., Leary, P.N. & Hart, M.B. (1990) Latest Cenomanian–earliest Turonian low-oxygen tolerant benthonic Foraminifera: a case study from the Sergipe basin (NE Brazil) and the western Anglo-Paris basin (southern England). *Palaeogeogr. Palaeoclim. Palaeoecol.* **77**, 145–177.

Morris, K.A. (1979) A classification of Jurassic marine shale sequences: an example from the Toarcian (Lower Jurassic) of Great Britain. *Palaeogeogr. Palaeoclim. Palaeoecol.* **26**, 117–126.

O'Brien, N.R. & Slatt, R.M. (1990) *Argillaceous Rock Atlas*. Springer-Verlag: New York.

Oschmann, W. (1988) Upper Kimmeridgian and Portlandian marine macrobenthic associations from southern England. *Facies* **18**, 49–82.

Pisciotto, K.A. & Garrison, R.E. (1981). Lithofacies and depositional environments of the Monterey Formation, California. In: *The Monterey Formation and Related Siliceous Rocks of California* (Eds R.E. Garrison, R.G. Douglas, K.E. Pisciotto, C.M. Isaacs & J.C. Ingle). Soc. Econ. Paleont. Miner. Pac. Sect. Spec. Publ. 15, 97–122.

Potter, P.E., Maynard, J.B. & Pryor, W.A. (1980) *Sedimentology of Shale.* Springer-Verlag: New York.

Pratt, L.M. (1984) Influence of paleoenvironmental factors on preservation of organic matter in middle Cretaceous Greenhorn Formation, Pueblo, Colorado. *Am. Ass. Petrol. Geol. Bull.* **68**, 1146–1159.

Raiswell, R. & Berner, R.A. (1985) Pyrite formation in euxinic and semi-euxinic sediments. *Am. J. Sci.* **285**, 710–724.

Rasmussen, H.W. (1977) Function and attachment of the stem of Isocrinidae and Pentacrinitidae: review and interpretation. *Lethaia* **10**, 51–57.

Rhoads, D.C. (1970) Mass properties, stability, and ecology of marine muds related to burrowing activity. In: *Trace Fossils* (Eds T.P. Crimes & J.C. Harper) pp. 391–406. Seel House Press: Liverpool.

Rhoads, D.C. (1974) Organism–sediment relations on the muddy sea floor. *Ann. Rev. Oceanogr. Mar. Biol.* **12**, 263–300.

Rhoads, D.C. & Morse, J.W. (1971) Evolutionary and ecologic significance of oxygen-deficient basins. *Lethaia* **4**, 413–428.

Rhoads, D.C. & Young, D.K. (1970) The influence of deposit-feeding organisms on sediment stability and community trophic structure. *J. Mar. Res.* **28**, 150–178.

Sageman, B.B. (1989) The benthic boundary biofacies model: Hartland Shale Member, Greenhorn Formation (Cenomanian), Western Interior, North America. *Palaeogeogr. Palaeoclim. Palaeoecol.* **74**, 87–110.

Savrda, C.E. & Bottjer, D.J. (1986) Trace fossil model for reconstruction of paleo-oxygenation in bottom-water. *Geology* **14**, 3–6.

Savrda, C.E. & Bottjer, D.J. (1987a) Trace fossils as indicators of bottom-water redox conditions in ancient marine environments. In: *New Concepts in the Use of Biogenic Sedimentary Structures for Paleoenvironmental Interpretation* (Ed. D.J. Bottjer). Soc. Econ. Paleont. Miner. Pac. Sect. Guidebook 52, 3–26.

Savrda, C.E. & Bottjer, D.J. (1987b) The exaerobic zone: a new oxygen-deficient marine biofacies. *Nature* **327**, 54–56.

Savrda, C.E. & Bottjer, D.J. (1989a) Trace fossil model for reconstructing oxygenation histories of ancient marine bottom waters: application to Upper Cretaceous Niobrara Formation, Colorado. *Palaeogeog. Palaeoclim. Palaeoecol.* **74**, 49–74.

Savrda, C.E. & Bottjer, D.J. (1989b) Anatomy and implications of bioturbated beds in 'black shale' sequences: examples from the Jurassic Posidonienschiefer (southern Germany). *Palaios* **4**, 330–342.

Savrda, C.E. & Bottjer, D.J. (1991) Oxygen-related biofacies in marine strata: an overview and update. In: *Modern and Ancient Continental Shelf Anoxia* (Eds R.V. Tyson & T.H. Pearson). *Geol. Soc. Spec. Publ.* **58**, 201–219.

Savrda, C.E., Bottjer, D.J. & Gorsline, D.S. (1984) Development of a comprehensive oxygen-deficient marine biofacies model: evidence from Santa Monica, San Pedro, and Santa Barbara basins, California continental borderland. *Am. Ass. Petrol. Geol. Bull.* **68**, 1179–1192.

Savrda, C.E., Bottjer, D.J. & Seilacher, A. (1991) Redox-related benthic events. In: *Cycles and Events in Stratigraphy* (Eds G. Einsele, W. Ricken & A. Seilacher) pp. 524–541. Springer-Verlag: Berlin.

Seilacher, A., Drozdewski, G. & Haude, R. (1968) Form and function of the stem in a pseudoplanktonic crinoid (*Seirocrinus*). *Palaeontology* **11**, 275–282.

Simms, M.J. (1986) Contrasting lifestyles in Lower Jurassic crinoids: a comparison of benthic and pseudopelagic Isocrinida. *Palaeontology* **29**, 475–493.

Stanley, S.M. (1970) Relation of shell form to life habits in the Bivalvia (Mollusca). *Geol. Soc. Am. Mem.* 125.

Surlyk, F. (1972) Morphological adaptations and population structures of the Danish Chalk brachiopods (Maastrichtian, Upper Cretaceous). *Det Kongelige Danske Videnskabernes Selskab, Biologiske Skrifter,* **19**, 1–57.

Thayer, C.W. (1974) Marine paleoecology in the Upper Devonian of New York. *Lethaia* **7**, 121–155.

Thayer, C.W. (1975) Morphologic adaptations of benthic invertebrates to soft substrata. *J. Mar. Res.* **33**, 177–189.

Thayer, C.W. (1983) Sediment-mediated biological disturbance and the evolution of marine benthos. In: *Biotic Interactions in Recent and Fossil Benthic Communities* (Eds M.J.S. Tevesz & P.L. McCall) pp. 479–648. Plenum Press: New York.

Thompson, J.B. & Newton, C.R. (1987) Ecological reinterpretation of the dysaerobic *Leiorhynchus* fauna: Upper Devonian Geneseo black shale, central New York. *Palaios* **2**, 274–281.

Wignall, P.B. (1990) Benthic palaeoecology of the Late Jurassic Kimmeridge Clay of England. *Palaeont. Ass. Spec. Pap. Palaeont.* **43**, 74 pp.

Wignall, P.B. & Myers, K.J. (1988) Interpreting benthic oxygen levels in mudrocks: a new approach. *Geology* **16**, 452–455.

Wignall, P.B. & Simms, M.J. (1990) Pseudoplankton. *Palaeontology* **33**, 359–378.

7   Hummocky cross-stratification

RICHARD J. CHEEL AND DALE A. LECKIE

Introduction

Hummocky cross-stratification (HCS) is a primary sedimentary structure that became popular during the late 1970s and early 1980s. Widespread recognition followed its description by Harms *et al.* (1975) although it had been earlier reported under different names (e.g. truncated wave-ripple laminae, Campbell, 1966; crazy bedding, Howard, 1971; truncated megaripples, Howard, 1972). The presence of HCS has since become a prime criterion for the recognition of ancient shallow-marine storm deposits; however, its reliability as an unequivocal criterion for this environment is now less certain. Despite the fact that HCS is widely accepted to be the product of waves, the structure continues to be the focus of ongoing debate regarding its mode of formation and, by implication, its specific palaeo-hydraulic interpretation. HCS in marine deposits has been variously attributed to formation by oscillatory flows produced by waves, combined oscillatory and unidirectional flows, and purely unidirectional flows. This diversity of hypotheses for HCS formation is justified because the basis for its interpretation is not as sound as that for many other sedimentary structures that are readily visible in modern settings or produced in laboratories. For example, our knowledge of the relationship between palaeohydraulics and stratification formed under unidirectional flows is quite advanced because we can easily dissect recognizable bedforms developed under known hydraulic conditions in rivers, intertidal areas and flumes, to precisely document the geometry of their internal structure. In contrast, HCS is found largely in ancient sediments and sedimentary rocks where palaeohydraulic conditions must be inferred from associated deposits and structures. Thus, HCS has been largely interpreted on the basis of inference rather than direct observation of the relationship between hydraulic processes and the form of the structure. The argument over the origin of HCS has recently been further complicated by the recognition of similar forms of stratification in settings in which wind-generated water-surface waves were an unlikely mechanism in its formation. The growing range of physical settings in which HCS-like stratification may have formed may justify the suggestion (Allen & Pound, 1985) that HCS has become just a 'bucket term' that embodies similar stratification styles which may be generated by a variety of processes or combinations of processes.

In this review we will concentrate on occurrences of HCS in unequivocally marine or lacustrine deposits where evidence points to an important influence by waves in its formation. We begin with a description of HCS and its stratigraphic/sedimentologic associations in ancient deposits and end with a summary of the various mechanisms that have been suggested for its formation. The aims of this review are: (i) to provide a basis for workers to recognize HCS in the field, and (ii) to indicate that there is no clear consensus for the mechanism(s) of HCS formation, although recent and ongoing work is leading to a better basis for its interpretation. Finally, we hope that this review will stimulate further work into the origin of this enigmatic sedimentary structure.

HCS — description and associations

Because of the paucity of modern and experimental examples of HCS, its palaeohydraulic interpretations must largely be based on its preserved characteristics in the geological record. The following description will review the basis for the recognition of HCS, including its common stratigraphic and sedimentologic associations, and stress characteristics that have led to the various ideas on HCS formation.

Characteristics of HCS

Grain size

The grain size of sediment in which HCS occurs varies from coarse silt to fine sand (Dott & Bourgeois, 1982; Brenchley, 1985; Swift *et al.*, 1987). HCS in coarser sediment is relatively rare but has been reported. Brenchley & Newall (1982) described HCS in sandstones with mean grain sizes ranging from 0.7 to 1.1 mm (i.e. up to coarse sand). Walker *et al.* (1983) noted that gravel may comprise beds displaying HCS. However, in cases where gravel-size sediment is present in HCS, the gravel normally makes up a small proportion of the total grain size distribution and is found largely as lag-deposits on surfaces within or at the base of beds displaying HCS. Therefore, the consensus seems to be that HCS is most common in very fine to fine-grained sand with the frequency of its occurrence decreasing dramatically with increasing grain size.

Morphology and geometry

Harms *et al.* (1975, p. 87) provided an early description of HCS (Fig. 7.1a) that has remained fundamental over the years (cf. Harms *et al.*, 1982). They pointed to four essential characteristics of individual cross-strata sets as being: (i) low-angle (generally less than 10° but up to 15°), erosional bounding surfaces; (ii) internal laminae that are approximately parallel to the lower bounding surface; (iii) individual internal laminae that vary systematically in thickness laterally and their angle of dip diminishes regularly; and (iv) dip directions of internal laminae and scoured surfaces are scat-tered (i.e. dipping with equal frequency in all directions). They also postulated that the stratification was due to deposition on a scoured bed surface characterized by low hummocks (bed highs) and swales (bed lows) with a spacing of one to a few metres and with a total relief of between 10 and 50 cm (e.g. Fig. 7.1). Hence, the form of the internal stratification was one of convex-upward hummocky laminae and concave-upward swaley laminae, essentially draped over the hummock and swale topography of the basal scoured surface (Fig. 7.1b).

Following Campbell (1966), Dott & Bourgeois (1982) employed a hierarchy of surfaces (Fig. 7.1b) to provide a careful description of HCS based on their observations. Here we employ their descriptive terms but details also come from other sources (Bourgeois, 1980; Hunter & Clifton, 1982; Brenchley, 1985).

First-order surfaces are surfaces of lithic change in discrete HCS beds (discussed below) and may bound HCS cosets or beds containing a sequence of various structures. The basal surface is commonly nearly flat and erosional, although local relief, up to several tens of centimetres (Fig. 7.3) may occur due to the presence of tool marks (scratches, grooves, prod-marks; Fig. 7.2d), gutter casts and/or rare flutes. This surface may be mantled with a lag of coarse debris of inorganic (intraclasts and/or extra-clasts) or biogenic origin (e.g. shell or bone material). In some instances, the upper surface is deeply scoured with a hummocky appearance (Fig. 7.4) whereas in many other cases this surface is rippled. Cosets of HCS range from decimetres to several metres in thickness, although the thickest cosets may actually consist of several amalgamated beds.

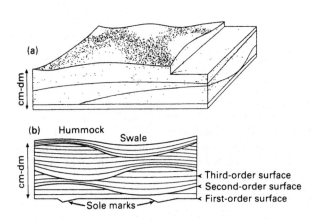

Fig. 7.1 (a) The original block diagram illustrating the characteristics of hummocky cross-stratification (HCS) (Harms *et al.*, 1975). (b) The form of stratification and first-, second- and third-order bounding surfaces commonly found in scour and drape hummocky cross-stratified sandstone beds.

(a)

(b)

(c)

(d)

Fig. 7.2 (a) Exhumed morphology of the feature (a bedform?) which produces HCS. (b) Wave ripples associated with the hummocky bed in (a). (c) Parting lineation on second-order surface exposed on same surface as hummocky bed in (a). The parting lineation is oriented orthogonally to the wave ripples shown in (b). (d) Sole marks from the bottom of a hummocky bed associated with that exposed in (a). The solemarks are parallel to the parting lineation in (c) and orthogonal to the wave-ripple crests in (b). From the Cretacous Gates Formation exposed near Hudson Hope, British Columbia. The relationships of the palaeocurrent data are explained in Leckie & Krystinik (1989).

Fig. 7.3 Hummocky cross-stratified sandstone bed showing eroded, upward-curving upper surface and erosional scour at base (arrow). Albian Gates Formation, Mt Spiker, British Columbia.

Fig. 7.4 A discrete HCS sandstone bed exhibiting a deeply-scoured upper hummocky surface. Notice, above the notebook the erosion of sub-parallel laminae. There is a thin, 1 cm layer of laminae of sandstone draping the exhumed feature.

Second-order surfaces are erosional surfaces within HCS cosets and are normally responsible for the stratification. They cut third-order surfaces but are contained by the first-order surfaces and therefore bound HCS sets (Fig. 7.1b). These surfaces commonly form the distinctly 'hummocky' surfaces in HCS that are characterized by laterally alternating synforms (swales) and antiforms (hummocks), although the antiforms are generally less common than the synforms. The relief on second-order surfaces ranges from several centimetres to approximately 50 cm (the same relief reported for hummocks and swales) with rare, extreme dip angles of

up to 35°. HCS set range from several centimetres up to 2 m in thickness although the latter extremes are probably the result of the amalgamation of beds. The erosional character of these surfaces may be obvious where relatively sharp, angular discordances occur between second- and third-order surfaces or may be subtle where third-order surfaces are nearly tangential with second-order surfaces (Fig. 7.5). The angular relationship commonly varies laterally, giving second-order surfaces the appearance of changing from discordant to concordant surfaces along an individual bed. The visibility of second-order surfaces is due largely to their

(b)

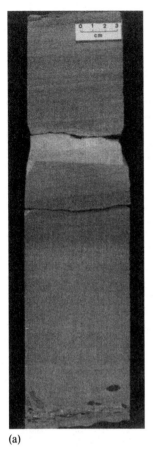

(a)

Fig. 7.5 Typical example of HCS inferred to be present in core. The third-order surfaces are nearly tangential with the second-order surfaces which makes interpretation as HCS difficult. Subtle examples such as this from the Western Canada Sedimentary Basin can be verified by association with nearby outcrop examples of correlative horizons. From the Turonian Cardium Formation, Alberta.

angular relationship with underlying strata. A change in grain size is typically not evident across second-order surfaces although they may be mantled by dispersed shale and/or siderite rip-up clasts, pebbles and shell debris. Such surfaces, exhumed in outcrop (Fig. 7.2a), typically display the hummock and swale topography described above. Most commonly, the plan form of hummocks and swales is circular, although elongate forms have also been reported (e.g. Handford, 1986). Exhumed second-order surfaces may display forms of parting lineation (Fig. 7.2c), including parting-step lineation (McBride & Yeakle, 1963) and current lineation (Allen, 1964).

Third-order surfaces bound individual laminae within HCS sets and account for many of the diagnostic characteristics of this structure although their visibility may depend on such fortuitous factors as the degree of weathering of the outcrop and cementation. Third-order surfaces are nearly concordant with underlying second-order surfaces which normally determine their overall geometry of internal laminae. Angles of dip are typically highest directly above erosional second-order surfaces but decrease upwards. Third-order surfaces are commonly mantled by mica, clay or comminuted plant debris (in many post-Silurian examples). Laminae defined by third-order surfaces tend to pinch and swell laterally and are most commonly thickest within swales, thinning upwards over hummocks. Individual laminae may or may not display internal grading, depending largely on the sorting of sand-size particles; well-sorted sand may not have a sufficiently wide range of grain sizes available to develop visible grading. For example, Bourgeois' (1980) observations of the Upper Cretaceous Cape Sebastian Sandstone of south-west Oregon reported centimetre-scale internal laminae in which grading was not detectable. However, Hunter & Clifton (1982), also working on the Cape Sebastian Sandstone, noted that, under certain conditions, light/dark couplets that comprise the internal laminae bore characteristics that suggested that they were normally graded. Like second-order surfaces, exhumed third-order surfaces may display various forms of parting lineation.

The scour and drape form of HCS is the most common variety of this structure although other forms have also been recognized, including vertical accretionary forms and migrating forms (Fig. 7.6). Several workers (e.g. Hunter & Clifton, 1982; Bourgeois, 1983; Brenchley, 1985; Allen & Underhill, 1989) described vertical accretionary HCS in which internal laminae thickened over hummocks rather than swales. As such, the hummocks appeared to have grown by accretion rather than formed by erosion of second-order surfaces. However, this 'accretionary' HCS was thought to be relatively rare. A variant of this form of HCS displays internal laminae that parallel the hummock and swale morphology of second-order surfaces (e.g. Allen & Underhill, 1989). Another type of HCS, described by Nøttvedt & Kreisa (1987) and Arnott & Southard (1990), is characterized by low-angle cross-strata sets (generally > 5 cm thick) filling shallow scours (swales) and which have a preferred, unimodal dip direction; hummocks are generally rare to absent. This latter structure has been termed 'low-angle trough cross-stratification' by Nøttvedt & Kreisa (1987) and 'anisotropic' HCS by Arnott & Southard (1990) in contrast to the more common *isotropic* HCS.

A structure that is closely related to HCS, and some say is a form of HCS (e.g. Dott & Bourgeois, 1983) is *swaley cross-stratification* (SCS; Fig. 7.7). SCS was initially introduced to describe 'a series of superimposed concave-upward shallow scours about 0.5–2 m wide and a few tens of centimeters deep' (Leckie & Walker, 1982, p. 143). SCS bears all of the essential characteristics of HCS summarized above with the following exceptions: the angle of dip of surfaces may be somewhat lower than in HCS; SCS lacks the convex-upward internal laminae of hummocks; it occurs in medium-grained or coarser sandstone more commonly than HCS; pebbles are not uncommon. This form of stratification may be characterized by nested sets of swaley stratification or isolated swales cutting horizontal to low-angle parallel lamination. SCS is not to be confused with trough cross-stratification, as the swales are generally not associated with angle-of-repose cross-strata. However, SCS may be traced laterally in outcrop into high-angle cross-stratification. Swales are wider and shallower than troughs and internal laminae may either drape lower margins of swales concordantly or pinch out discordantly against the margins of swales.

The adherence by later workers to the essentials of the definition of HCS (given by Harms *et al.*, 1975) above has been a matter for some debate. For example, Brenchley (1985) reported slopes on hummocky surfaces up to 35° and spacings as small as the spacing of wave ripples (i.e. centimetres). These deviations from the original definition have led some to suggest that these smaller forms (commonly termed micro-HCS) are not true HCS as Harm *et al.*

Forms of HCS in shallow-marine sandstones

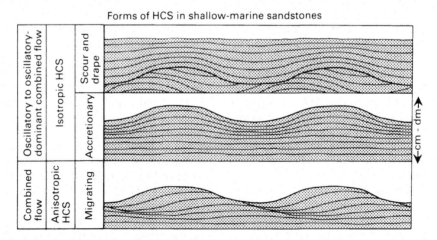

Fig. 7.6 Forms of HCS in shallow-marine sandstones. Unidirectional component of combined flow forming anisotropic HCS is from right to left.

Fig. 7.7 Swaley cross-stratified sandstone in medium-grained sandstone of the upper shoreface. From the Santonian Chungo Member, Wapiabi Formation in Alberta. Scale is 15 cm.

(1975) had defined it. For example, Duke (1987, p. 345) argued that HCS-like stratification with hummock spacings below the '1 m lower limit assigned to HCS by Harms *et al.* (1975)' is not true HCS. However, the first experimentally-produced analogues of HCS identified by Harms *et al.* (1982) had spacings on the order of a couple of decimetres (and Harms *et al.*, 1982, point out the discrepancy with HCS observed in the field). The scale of HCS may or may not be a limiting factor in defining HCS but may represent a natural variation in scale that reflects the breadth of conditions over which this structure may form. Sherman & Greenwood (1989, p. 985) emphatically state that 'there is no apparent physical rationale for the 1 m lower limit on hummocky cross-stratification wavelength' and Campbell (1966) suggested that hummocks may occur with wavelengths of 0.1–10 m. Alternatively, many of the examples of HCS that differ significantly from the form defined above may represent similar stratification styles that formed by different processes. The breadth of variation in form of HCS that has grown in the literature may be partly responsible for the similar proliferation of ideas regarding its mode of formation that is discussed below.

Grain fabric of HCS sandstones

Very little work has been directed at the determination of fabric of sandstones displaying HCS. Handford (1986) described the orientation of macroscopic fossils associated with HCS and found a preferred long-axis alignment, essentially parallel to elongate crests of hummocks. A detailed study of grain fabric was reported by Cheel (1991) based on samples oriented with respect to sole marks on the base of discrete HCS sandstone beds interbedded with shale. Cheel found that particle a-axes, measured in plan view, varied widely but displayed modes oriented approximately normal to sole marks and parallel to the associated ripple crest. However, he pointed out that HCS displaying parting lineation oriented parallel to sole marks (as is commonly observed; e.g. Fig. 7.2) should display grain long axes similarly aligned parallel to sole marks. This suggests that a-axis alignment of grains in HCS may be quite variable. Imbrication of grains measured in the vertical plane through HSC sandstone varied about a mean of 0°, parallel to visible lamination. In some cases, this variation in imbrication was markedly cyclic about the 0° mean. This pattern of imbrication was interpreted in terms of the action of symmetrically oscillating currents during the formation of isotropic HCS. In contrast, in the basal parallel-laminated interval of an HCS bed, the mean particle imbrication was approximately 13° into the flow direction (based on the sole marks) and varied quasi-cyclically about that mean. His interpretation of the style of imbrication in basal parallel-laminated sandstone envisaged deposition under a combined flow with a net offshore-directed unidirectional component.

HCS associations

With the onset of widespread recognition of HCS, several studies reported its occurrence in particular associations with other structures (e.g. Hamblin & Walker, 1979; Dott & Bourgeois, 1982; Brenchley, 1985). The most common occurrences of HCS in the ancient record can be classified into two such associations: (i) discrete sandstone beds interbedded with mudstone; and (ii) amalgamated sandstones; however, specific associations vary widely in nature (cf. Dott & Bourgeois, 1983).

Discrete HCS sandstones

Dott & Bourgeois (1982) were the first to propose an 'ideal sequence' or model showing structures that are preferentially associated in outcrop within sandstone beds containing HCS; an ideal sequence that began a lineage of such sequences for HCS sandstones. This sequence consisted of sharp-based sandstone interbedded with bioturbated mudstone; the basal-scour surface includes sole marks and is mantled by a lag of coarse debris overlain by an interval of HCS passing upward into flat lamination and ultimately to cross-laminae associated with symmetrical ripple forms that cap the sandstone. A similar sequence of associations was proposed by Walker et al. (1983) which differed in detail from that proposed by Dott & Bourgeois (1982, 1983) by the occurrence of a basal parallel (horizontal to subhorizontal) laminated interval directly overlying the basal erosional surface. With this modification, Walker et al. (1983) emphasized an analogy with the Bouma turbidite sequence. The evolution of the model continued as more observations were made and data collected. For example, Fig. 7.8 (from Leckie & Krystinik, 1989) shows a recent version of the early ideal sequences and contains considerably more information and more variability than the earlier sequences; the new information includes the occurrence of parting lineation on surfaces within HCS, palaeocurrent relationships and a range of ripple types capping the beds, from purely wave-formed ripples through to purely current ripples. The trend of the parting lineation is generally orthogonal to wave-ripple crests and subparallel to sole marks at the bases of hummocky beds (e.g. Fig. 7.2; Brenchley, 1985; Leckie & Krystinik, 1989). Rarely, a polymodal trend of parting lineation has been observed on second- or third-order surfaces (D.A. Leckie & L.F. Krystinik, unpublished

observations). Capping wave-ripples are typically straight-crested with bifurcating patterns, although irregular forms are not uncommon, including polygonal, ladderback and box patterns. In addition, Leckie & Krystinik (1989) include directional relationships between structures in the HCS beds with regional shoreline and palaeoslope (Fig. 7.8). Specifically, they showed that directional structures, such as sole marks and parting lineation, indicate palaeoflows directed offshore, orthogonal to regional shoreline-trend indicators. Similarly, capping ripples have crests aligned approximately parallel to regional shoreline and the internal cross-laminae, when present, indicate migration offshore. Such relationships had been suggested earlier on the basis of local studies (e.g. Hamblin & Walker, 1979; Brenchley, 1985, Rosenthal & Walker, 1978) but data provided by Leckie & Krystinik (1989) suggested that the directional associations may be the norm for discrete HCS sandstones.

Stratigraphic associations with discrete HCS beds include fine-grained sandstone beds characterized by sharp bases, with sole marks oriented similarly to those on HCS beds, and horizontally-laminated and current-rippled intervals that are markedly similar to B–C (division) Bouma turbidites. Such beds typically occur stratigraphically beneath the discrete HCS sandstones in progradational sequences (e.g. Hamblin & Walker, 1979: Leckie & Walker, 1982).

Amalgamated HCS sandstones

This association of HCS is characterized by thick (up to several tens of metres) sandstone and differs from the other association by the lack of mudstone (except as local lenses) and the absence of a preferred sequence of structures. Amalgamated HCS commonly occur above the discrete HCS beds in regressive shoreline successions (e.g. Hamblin & Walker, 1979; Leckie & Walker, 1982) and is representative of sedimentation in the lower shoreface. First-order surfaces may be recognized within amalgamated sandstone beds by the presence of a lag or where they overlie intensely bioturbated horizons (Fig. 7.9), discontinuous mudstone beds or concentrations of mica and fine plant debris.

Swaley cross-stratification, as originally defined by Leckie & Walker (1982), is not the amalgamated HCS as described here, although there is a growing tendency amongst some authors to state this. For example, Dott & Bourgeois (1983), McCrory &

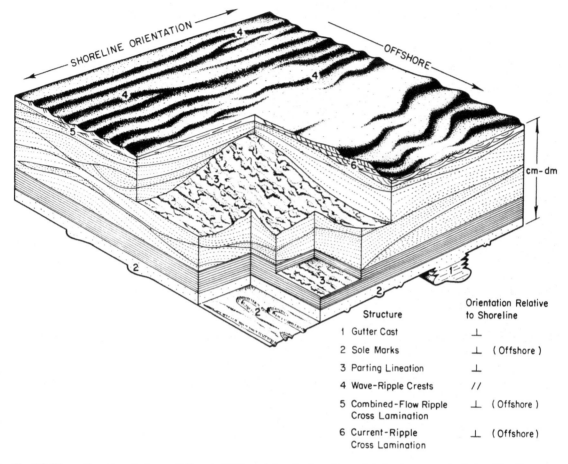

Structure	Orientation Relative to Shoreline
1 Gutter Cast	⊥
2 Sole Marks	⊥ (Offshore)
3 Parting Lineation	⊥
4 Wave-Ripple Crests	//
5 Combined-Flow Ripple Cross Lamination	⊥ (Offshore)
6 Current-Ripple Cross Lamination	⊥ (Offshore)

Fig. 7.8 The sedimentary features of discrete HCS sandstone beds and associated palaeocurrent relationships (from Leckie & Krystinik, 1989).

Walker (1986) and Plint & Walker (1987) suggest that the swaley cross-stratification is typical of amalgamated sandstone beds. Duke (1985, p. 171), however, specifically stated that swaley cross-stratified sandstones do not show evidence of amalgamation. In a vertical, progradational succession, discrete HSC is overlain by amalgamated HCS which, in turn, is overlain by SCS.

Proximal–distal relationships

Only a few studies have been directed at outlining the lateral variability of HCS beds (e.g. Dott & Bourgeois, 1982; Brenchley, 1985). From these, the following generalizations regarding the change in HCS-bearing sandstones from inferred proximal to distal settings can be made:

1 Thicker sandstone beds occur in a proximal setting; amalgamated HCS sandstone grades laterally into discrete HCS beds in more distal deposits.
2 Decreasing grain size.
3 Decreasing sandstone to shale ratio.
4 Increasing depth and steepness of scour at the base of sandstone beds (i.e. more prominent gutter casts).
5 Increased thickness of the parallel-laminated interval above the erosional base of discrete HCS storm beds (although thick units of parallel-laminated sandstone may predominate in more proximal settings).
6 An increase in the abundance of wave ripples and eventually current ripples, and, more distally, graded beds.
7 Better-developed normal grading within indi-

(a)

(b)

Fig. 7.9 (a) Amalgamated HCS sandstone beds, distinguished by intensely bioturbated sediment. (b) Sharp-based sandstone with vertical escape tubes. From the Albian Gates Formation, Mt Spieker, British Columbia.

vidual beds occurs more distally.

8 An increase in the intensity of bioturbation, particularly at the tops of HCS beds.

Trace fossils

The bioturbation associated with progradational successions containing HCS sandstones typically contain elements of the *Skolithos* ichnofacies or a mixed *Skolithos–Cruziana* ichnofacies (Fig. 7.10; Frey, 1990). The *Skolithos* ichnofacies is character-

istic of high-energy settings with a well-sorted substrate (Seilacher, 1967). The organisms are typically infaunal suspension feeders, with long vertical burrows, commonly with reinforced wall linings. Representative trace fossils include *Skolithos, Ophiomorpha, Diplocraterion, Monocraterion,* and *Arenicolites* occurring in a low diversity but relatively high density (Frey & Pemberton, 1984). This *Skolithos* ichnofacies represents colonization by opportunistic or resilient organisms, of a substrate which has been rapidly deposited. After emplace-

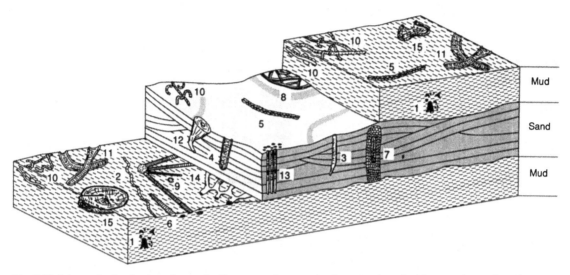

Fig. 7.10 Schematic distribution of trace fossils commonly occurring in storm-deposited hummocky sands and bounding muds. 1, *Chondrites;* 2, *Cochlichnus;* 3, *Cylindrichnus;* 4, *Diplocraterion;* 5, *Gryochorte;* 6, *Muensteria;* 7, *Ophiomorpha;* 8, *Palaeophycus;* 9, *Phoebichnus;* 10, *Planolites;* 11, *Rhizocorallium;* 12, *Rosselia;* 13, *Skolithos;* 14, *Thalassinoides;* 15, *Zoophycos.* (From Ekdale *et al.,* 1984.)

ment of the sand, the original resident community or a subsequent one may overprint the *Skolithos* ichnofacies resulting in a mixed *Skolithos–Cruziana* ichnofacies. Vertical escape burrows (Fig. 7.9b) represent the traces of organisms suddenly buried by sand in an attempt to rise to the new substrate following a storm.

In proximal settings, HCS beds may be interbedded with shale and sandy-shale layers containing the *Cruziana* ichnofacies, characterized by the trace fossils *Cruziana, Dimorphichnus, Diplichnites, Teichichnus, Asteriacites, Phycodes, Rosselia, Scolicia, Asterosoma, Thalassinoides* and horizontal *Ophiomorpha*, representative of a stable benthic community (Ekdale *et al.* 1984; Frey, 1990). In amalgamated HCS sandstones, evidence of burrowing by colonizing organisms may be removed by subsequent erosion which gives the sediment an unbioturbated aspect (Vossler & Pemberton, 1989). In a distal setting the *Cruziana* ichnofacies may dominate the sandstones and mudstones.

Coarse-grained ripples

Leckie (1988) noted that conglomeratic beds capped with large-scale (metre spacing and decimetre amplitude) wave ripples, termed coarse-grained ripples, occur in some shallow-marine suc-cessions where they appear to be stratigraphically equivalent to HCS sandstones. In fact, in some outcrops the conglomerates can be traced laterally into HCS sandstones. Leckie argued that coarse-grained ripples replaced HCS in coarse sediment, suggesting that both formed in response to the same hydraulic processes. Cheel & Leckie (1991) have further demonstrated that these conglomerates form preferred sequences that include intervals comparable to the sequences typical of discrete HCS sandstones (Fig. 7.11). The conglomeratic beds are sharp-based, with rare offshore-oriented sole marks, overlain by onshore-imbricate particles, including shale and/or sandstone intraclasts, passing upwards into low-angle cross-stratification, associated with capping symmetrical ripples, and on which particles dip onshore. Cheel & Leckie (1991) interpret this sequence as the product of offshore-directed currents which transport gravel and sand away from the shoreline where it is deposited and subsequently reworked by onshore-directed, asymmetrical oscillatory currents induced by shoaling swell waves.

Where does HCS form?

Most HCS is preserved in shallow-marine suc-cessions, in deposits laid down anywhere from

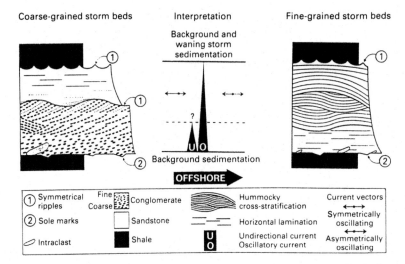

Fig. 7.11 Comparison of the vertical succession of structures in coarse-grained storm beds and fine-grained storm beds.

intertidal or shallow subtidal to outer-shelf environments. Leckie & Walker (1982) and Brenchley (1985) argued for the deposition of HCS sandstones as far offshore as 100 km. On modern shelves, sand beds that may be analogous to HCS sandstones have been observed 24 km offshore in the Gulf of Mexico following the passage of Hurricane Carla (Hayes, 1967). Gadow & Reineck (1969) reported 10-cm-thick laminated sand and mud couplets deposited 50 km offshore in 14–24 m water depths by winter storms on the German Bight. Nelson (1982) indicated that sandy-storm beds may be deposited up to 100 km offshore in the modern Bering Strait. Coarse-grained ripples, which may form under hydraulically similar conditions to HCS (Fig. 7.11), have been observed on the modern shelf to depths of 160 m. The association of the trace fossil *Zoophycos* within HCS sandstones also gives a clue to the maximum depths at which HCS forms. During the Cretaceous, *Zoophycos* was probably formed by organisms that inhabited a shelf setting (Bottjer *et al.*, 1984) at water depths up to approximately 180 m (Tillman, 1985).

Greenwood & Sherman (1986) have described HCS formed in the surf zone (water depths of less than 2 m) from Lake Huron, Canada. This is consistent with observations from interglacial, glacial and post-glacial deposits around Lake Ontario, Canada, where Eyles & Clark (1986) observed HCS, SCS and coarse-grained ripples (reconstructed wave conditions of less than 20 m water depth, wave heights of 5 m and periods of 5 s). HCS formed in

an ancient lacustrine setting has also been described (e.g. Fielding, 1989). Allen (1985a) illustrated coarse-grained ripples formed (see Allen's book cover) in pebbly, shelly, very coarse sands in the intertidal zone of Guernsey. Therefore, hummocky cross-stratified sands may similarly be expected to form during storms at high tide in the intertidal zone on coasts having fetch of as little as a few tens of kilometres.

Thus, in summary, it appears that HCS can be expected to form in water depths ranging from the intertidal zone to the outer shelf at nearly 200 m depth. The structure may form in close proximity (metres) to the shoreline or be as far as 100 km out to sea. HCS can be expected in shallow-marine deposits and those of large bays or lakes. The preservation potential of HCS, though, appears to be very limited in extremely shallow depths as most ancient examples are from deeper-water, shallow-marine settings.

How does HCS form?

The question of what physical processes lead to the formation of HCS has been the subject of considerable debate. The majority of HCS is preserved in deposits of shallow-marine or lacustrine environments where waves, unidirectional or combined flows may have been effective. Most early workers consistently attributed HCS formation to waves (Campbell, 1966; Howard, 1972; Harms *et al.*, 1975). For example, while Harms *et al.* (1975) were not certain of the exact process of HCS formation, they were explicit in citing powerful, wind-induced, surface-gravity waves, generated during storms, as the dominant mechanism in its formation.

As more information became known about HCS in shallow-marine deposits, and particularly its sedimentological associations, the role of waves became less clear and an auxiliary question became central to the discussion; how are the sands in which HCS occurs transported to their site of deposition on the shelf? The mode of transport of these sands is almost certainly as suspension load (Dott & Bourgeois, 1982); the predominantly very fine- to fine-grained sands, in which HCS occurs, are prone to suspension transport under shear velocities that differ little from the critical shear velocity required to initiate motion of such grain sizes (Blatt *et al.*, 1980, Fig. 4–6, p. 103). Unidirectional flows are required to transport the sands to great distances offshore; therefore, the possible role of these currents has to be considered in the formation of HCS. Furthermore, solemarks and coarse lags at the base of HCS beds suggest episodic and rapid input of sands by powerful currents. There are currently two schools of thought on the mechanisms for the transport of sand out onto the shelf: (i) those who attribute transport to a storm-surge relaxation/turbidity current mechanism, based on observations from the geological record; and (ii) those who attribute transport to wind- and barometric pressure-driven mechanisms, based largely on observations on modern continental shelves.

Transport onto the shelf

Storm-surge relaxation/turbidity current mechanism

A commonly invoked interpretation, based primarily on the similarities between discrete HCS storm beds and Bouma turbidites, is that HCS sands are transported offshore by powerful, storm-generated, offshore-directed turbidity currents (Fig. 7.12a). A storm-surge ebb origin for these turbidity currents has been proposed, based largely on Hayes' (1967) interpretation of a graded bed in the Gulf of Mexico, originating from Hurricane Carla, in 1961, possibly one of the largest recorded tropical storms (Colon, 1966). Hayes (1967) inferred that storm-surge ebb (relaxation) occurred after landfall of the storm and that sediment was transported seaward through previous washover channels. Hayes stated

that transport across the shelf occurred as a density current. The standard interpretation is that during the passage of a storm, water builds up against the shoreline (i.e. coastal set-up) because of wind drag from onshore winds and lowered barometric pressure, causing a local and temporary rise in sea-level. The storm-surge build up for Hurricane Carla was between 4 and 6 m (Hayes, 1967) and in the Bay of Bengal storm surges have been reported up to 7 m (Flierl & Robinson, 1972). Nearshore sediment taken into suspension by turbulence is transported seaward by offshore-directed, bottom-flowing return currents initially driven by the coastal set-up. These currents become particularly intense when the coastal set-up collapses as the storm abates. Because of the relatively high density of the sediment-charged flows compared to ambient seawater, the bottom currents act as underflows transporting sand seaward across the shoreface and shelf, in response to the momentum afforded by storm-surge relaxation and gravity acting on the density current. These currents have subsequently been called turbidity currents, storm-surge ebb currents, storm-rip currents and bottom-hugging storm currents. Once the sediment is transported onto the shelf it is deposited as the current decelerates, losing capacity and competence, at depths that continue to be influenced by waves produced by the same storm that caused suspension in the nearshore.

The above scenario prompted several objections including the question of whether turbidity currents are a viable transport mechanism on the relatively low shelf slopes. This led Walker (1984, 1985) to suggest that during coastal set-up of water, the sandy nearshore substrate may become liquefied to a depth of several metres by storm waves, causing it to slump seaward; the slump would accelerate downslope and develop into a turbidity current. However, Hayes' (1967) interpretation of the deposits of Hurricane Carla was questioned by Morton (1981) who re-evaluated data on shelf topography, geomorphology and storm currents associated with that storm, concluding that the storm deposits were the result of wind-driven currents and not a powerful storm-surge return current. In fact, following a re-examination of Hurricane Carla data and modelling of storm flows in general, Snedden *et al.* (1988) concluded that most sediment transport occurred prior to storm landfall and not following it, as proposed by Hayes (1967).

Using the turbidity current scenario for sand emplacement, discrete HCS beds have been inter-

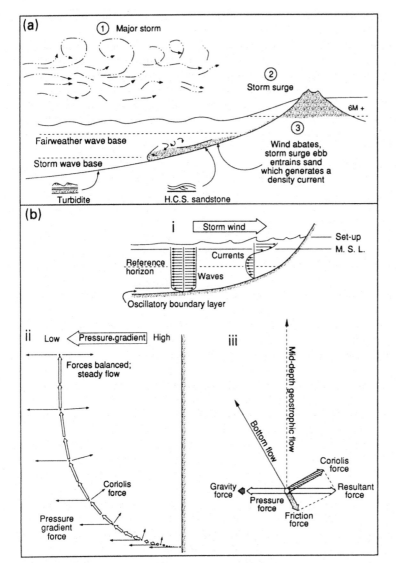

Fig. 7.12 (a) The turbidite model initially presented to account for the transport and deposition of HCS beds (after Hamblin & Walker, 1979). (b) Geostrophic flow mechanism to explain the transport of sediment on to the shelf (from Duke, 1990).

preted as the deposits of waning flows in a manner similar to the interpretation of Bouma turbidites. The basal erosional surfaces, exhibiting directional sole marks, are formed while the turbidity current is active, and the overlying horizontally-laminated interval is commonly attributed to deposition while the unidirectional flow wanes. With cessation of the unidirectional flow, sediment is reworked by the powerful storm waves at depths above storm wave-base. Capping wave ripples form while surface waves diminish as the storm wanes. Further off-shore, below storm wave-base, deposition of sediment continues from the unidirectional, waning

turbidity current, resulting in a sharp-based graded beds virtually identical to Bouma turbidites.

Estimates of the recurrence interval of storms capable of forming storm beds have been calculated by dividing the number of storm beds in a sedimentary succession by the duration of that succession. Although many errors are inherent in this method, there is remarkable consistency in the results of different workers. In general, the storm beds have a recurrence frequency of a few thousand years (Table 7.1). As such, it has been stated that HCS beds may be the result of catastrophic storms, the likes of which have not yet been recorded. This is in marked

Table 7.1 Recurrence intervals of emplacement events for HCS sandstones

Age	Recurrence interval (years)	References
Kimmeridgian	3200–4000	Hamblin & Walker, 1979
		Walker, 1985
Devonian	400–2000	Goldring & Langenstrassen, 1979
Ordovician	10 000–15 000	Brenchley et al., 1979
Triassic	2500–5000 or 5000–10 000	Aigner, 1982
Ordovician	1200–3100	Kreisa, 1981

contrast to observations on modern mid-latitude storm-dominated shelves where 3–5 sediment transport events can occur per year (Swift *et al.*, 1986).

Wind-forced current (geostrophic flows)

Oceanographers have objected to the turbidity-current transport mechanism on the grounds that the typical slope of modern shelves is insufficient to initiate turbidity currents and is certainly inadequate to induce autosuspension that is required to sustain such currents (Pantin, 1979; Parker, 1982). Furthermore, the flows like those envisaged by geologists have never been observed flowing orthogonally offshore in modern coastal settings. Rather, observations of storm-generated currents on modern shelves show that downwelling currents resulting from coastal set-up during a storm are deflected by Coriolis force (to the right in the northern hemisphere and to the left in the southern hemisphere) (Fig. 7.12b). The net result is that most of the water column is set in motion flowing more or less parallel to bathymetric contours and local shorelines. There is considerable ambiguity regarding the landward limit to which geostrophic currents will affect the shoreface. Swift & Niedoroda (1985) suggested that geostrophic veering should take place in 20–40 m water depth whereas Swift *et al.* (1987) indicated water depths as shallow as 15 m.

Yet, in spite of theoretical modelling and modern observations supporting the occurrence of these geostrophic flows, the geostrophic transport mechanism is not consistent with directional data from the ancient record (Leckie & Krystinik, 1989).

Many workers of the ancient record continued to doubt that geostrophically-balanced flows were involved in the emplacement of HCS beds. The reluctance was based on the fact that palaeocurrent indicators associated with HCS beds showed that the flows were directed offshore, not alongshore. In an attempt to reconcile the apparently conflicting data, Duke (1990) and Duke *et al.* (1991) suggested that the reported palaeocurrent patterns could be explained by combined flow produced by the interaction of a nearly shore-parallel geostrophic current and oscillatory fluid motion at the bed produced by water-surface gravity waves propagating directly onshore (i.e. wave crests parallel to bathymetric contours and shoreline). Under such combined flows the bottom feels the unidirectional current directed approximately obliquely offshore and the oscillatory current directed alternately onshore and offshore. However, the offshore stroke of the oscillatory component combines with the slightly-offshore unidirectional component to exert maximum boundary shear stress in the offshore direction. Hence, large objects such as shell material and pebbles on the bed will most likely move in the direction of maximum shear stress, producing structures aligned parallel to the direction of wave propagation; orthogonal to the shoreline. Figure 7.13 shows their model (Duke, 1990; Duke *et al.*, 1991) for the emplacement of discrete HCS beds under waning, geostrophically-balanced flows. This model is superficially similar to that for storm beds emplaced by turbidity currents except that the unidirectional component of the current is derived from the offshore-directed component of a shoreline-oblique, geostrophically-balanced combined flow. In this model, basal sole marks are formed and basal laminated sands are deposited while the unidirectional component of the flow is significant. However, HCS forms when the unidirectional component has waned sufficiently so that the boundary is influenced by a strongly-oscillatory dominant or pure-oscillatory flow with wave ripples forming as the storm waves subside.

Mode of formation

Because HCS has not actually been observed to form, ideas on its mode of formation are inferred from its sedimentological relationships and associations. The form of HCS is highly variable, suggesting that there may be several possible types and possibly different modes of formation. As noted

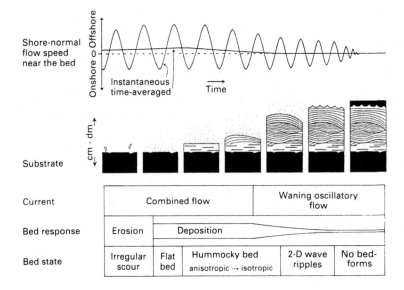

Shore-normal flow speed near the bed

Fig. 7.13 A model for the formation of discrete HCS storm beds by geostrophically balanced flows during waning storm conditions (after Cheel, 1991; Duke *et al.*, 1991).

above, there are at least three variants of this structure: scour and drape; vertical accretion (both isotropic forms); and low-angle migratory forms (anisotropic HCS). Figure 7.6 illustrates these forms of HCS and indicates the types of currents that are most commonly thought to form them in a shallow-marine setting.

The most common variant of HCS, the scour and drape form, appears to be produced by scour events that form an *erosional* hummocky topography (i.e. second-order surfaces) followed by periods of deposition in which sediment falls out from suspension, draping the scoured surface and subduing its initial relief. Repeated scour- and drape-form HCS are characterized by thickest accumulation in swales, thinning over the hummocks. Such HCS appears to be unrelated to stable bedforms on the substrate. Many workers attribute the currents that form the scour and drape topography to powerful oscillatory flows (e.g. Harms *et al.*, 1975, 1982). One possible scenario for the formation of HCS by oscillatory currents under waves that has been little improved upon was given by Dott & Bourgeois (1982). They suggested that deposition of sand from suspension formed HCS under flow conditions, straddling the hydraulic stability fields for ripples and wave-induced sheet flow, that moulded a gently undulating (hummocky) bed onto which sediment was draped. Internal laminae form as sediment drapes the surface and third-order surfaces are formed due to grain sorting through an individual

wave cycle or some cyclic variation in the current with the passage of a single wave train. Internal second-order surfaces were thought to form in response to fluctuations in intensity of the current acting on the boundary causing local scour during a period of overall net deposition onto the bed.

The scour and drape form of HCS might be best explained by the mechanisms suggested by Dott & Bourgeois (1982) with some modification. The fact that internal laminae drape and diminish the hummocky relief suggests that a hummocky bedform is not stable under conditions that form the structure but are only stable during periods of intense flow that causes erosion of the hummocky surface. Such erosion during a storm might occur due to the temporary development of constructive waves on the water surface. Swift *et al.* (1983) described the generation of thick clouds of sediment that rose off the bottom during a storm with the passage of groups of exceptionally high waves formed by constructive interference. The formation of these clouds must involve the local addition of sediment into suspension by erosion that might form hummocky second-order surfaces. Between periods of wave construction the 'normal storm waves' would act as sediment continued to deposit, causing the temporal variation in flow strength that results in the formation of internal laminae. The action of lesser waves than those during constructive interference might also explain the rare occurrence of wave ripples within HCS sets (e.g. Craft & Bridge, 1987), possibly formed under aggradation rates that were low enough to allow the development of the relatively small-scale bedforms. This scenario differs from Dott & Bourgeois (1982) in that the hummocky surface is not due to the generation of a stable bedform but is inherited from the form of an erosional surface: a surface that might mimic the morphology of the stable bedform under the same conditions but with net deposition on the bed. Subsequent deposition of sediment onto the hummocky erosional surface acts to bring the topography into equilibrium with the normal storm-wave conditions (and this equilibrium bed appears to be more-or-less flat). Hence, the currents that produce the hummocky form are associated with erosion, during the most intense conditions on the bed.

Vertical accretion forms (Fig. 7.6), characterized by thickest laminae within hummocks, appear to involve the growth of a *depositional* hummocky topography. This type of HCS might represent the product of currents, possibly produced by sustained constructional waves that are less transient than the case of scour and drape. Under conditions of rapid deposition, such sustained currents, capable of building a stationary hummocky bedform that is in equilibrium with the prolonged, oscillatory currents generated by the constructional waves, would form vertical accretionary sets. Hummocky bedforms that would produce such stratification were produced under purely oscillatory flows and strongly oscillatory combined flows by Arnott & Southard (1990). In this case, a true bedform is constructed whereas in the case of scour and drape HCS, only an erosional surface that mimics the bedform develops under conditions of net aggradation.

The low-angle migratory forms (Fig. 7.6) may represent similar bedforms that form accretionary HCS but which migrate over the substrate, preferentially preserving internal laminae that dip in the direction of migration (Nøttvedt & Kreisa, 1987). Arnott & Southard (1990) produced stationary and migrating bedforms experimentally in a wave duct. The stationary forms were stable under oscillatory flows and strongly oscillatory-dominant combined flows. However, migrating forms developed when the velocity of the unidirectional component exceeded a few percent of the orbital velocity of the oscillatory component.

Several other proposed mechanisms of HCS formation stress the role of unidirectional currents in its formation. Swift *et al.* (1983) reported HCS on modern shelves in 15–40 m water depth. Their examples of HCS described from box cores were formed by wave-modification of megaripples, as seen on side-scan sonar images, that were stable during storm-generated geostrophic flows. Box cores taken from the bedforms displayed wedge-shaped sets of heavy mineral-rich laminae in fine to very fine-grained sands. Swift *et al.* (1983) explained the wedge-shaped sets as curved lamina intersections and likened the structure to HCS. Hence, the HCS was attributed to combined flows, and specifically to wave modification of what would have otherwise been megaripples under unidirectional flows. Elsewhere, Greenwood & Sherman (1986) suggested that, in a lacustrine setting, a unidirectional flow (in their case, a longshore current) was crucial to the formation of HCS. They argued that without the combined-flow component, purely oscillatory waves would only produce flat-bed conditions. Similarly, Allen (1985b) argued that oscillatory currents alone could not form hummocky bedforms of metre-scale wavelength, the development of which required a unidirectional current. However, the results of flume experiments by Arnott & Southard (1990), shed some doubt on this conclusion.

Figure 7.6 classifies the three forms of HCS described in this review according to the morphology and geometry of internal cross-strata and (in a currently speculative manner) in terms of the formative flows. Isotropic and anisotropic HCS may well record a continuum of flow types from purely oscillatory currents which form stationary bedforms to combined flows in which the undirectional component causes bedform migration. In fact, taking this continuum to completion, if isotropic HCS is the oscillatory current end member, then the opposite end member, formed under unidirectional dominant combined flows and purely unidirectional flows, will be trough cross-stratification. This full continuum is suggested for large bedforms (i.e. here we do not consider small wave, combined flow and current ripples) by the results of Myrow & Southard (1991). With sufficiently high rates of aggradation on the bed the isotropic and anisotropic forms, respectively, will become vertically and obliquely (downcurrent) climbing forms of aggradational HCS. Duke *et al.* (1991) also implied such a continuum from isotropic to anisotropic HCS. It is possible that close inspection of the basal portions of HCS storm beds, particularly through the transition from the horizontally laminated interval to the

HCS interval of HCS storm beds, may show that anisotropic HCS is more common in shallow-marine deposits than we currently believe.

Conclusions

We are left with the ongoing dilemma of the question of how does HCS form: oscillatory currents or combined flows in which the unidirectional component is crucial to the formation of HCS. A wealth of evidence points to the importance of waves in forming isotropic HCS, but the existence of a unidirectional current associated at least with the emplacement of the sands is also indisputable. Limited fabric data (Cheel, 1991) and experimental work (Arnott & Southard, 1990) indicate that it is the oscillatory currents that form isotropic HCS. However, anisotropic HCS is sensibly interpreted to be the product of a combined flow with an effective unidirectional component that causes the bedform to migrate. Thus, the variety of forms of HCS appear to represent a continuum of flow types from purely oscillatory to combined flows with specific form of internal strata being additionally controlled by the rate of bed aggradation (Fig. 7.6).

In this review we have limited our discussion to HCS formed in shallow-marine and lacustrine settings. Recent literature has also described a similar structure in settings where waves are not a likely mechanism for its formation. For example, Prave & Duke (1990) described micro-HCS from turbidites deposited in a bathyal setting. They convincingly argued that the stratification was formed by antidunes produced in a unidirectional turbidity underflow. Similarly, Rust & Gibling (1990) described three-dimensional antidunes formed in fluvial sediments which closely resemble aggradational HCS. Therefore, while the evidence for HCS formation in shallow-marine settings by wave-induced oscillatory flows is growing stronger, we are now beginning to see evidence for other mechanisms of formation of HCS-like stratification. Hopefully, the next phase of research into this form of stratification will include the development of new criteria and techniques for distinguishing the variety of similar structures, all of which are now termed HCS.

Acknowledgements

R.J. Cheel gratefully acknowledges ongoing funding in the form of Operating Grants from the Natural Sciences and Engineering Research Council of Canada. The authors thank Tony Hamblin for his comments on an earlier version of the manuscript and also P.A. Allen and R. Goldring for their reviews of this paper. We also thank Mike Lozon (Brock University) for drafting several of the figures in this chapter. This chapter is Geological Survey of Canada Contribution No.27991.

References

Aigner, T. (1982) Calcareous tempestites: storm-dominated stratification in Upper Muschelkalk limestones (Middle Trias, SW Germany). In: *Cyclic and Event Stratification* (Eds G. Einsele & A. Seilacher) pp. 180–198. Springer-Verlag: Berlin.

Allen, J.R.L. (1964) Primary current lineation in the Lower Old Red Sandstone (Devonian), Anglo-Welsh basin. *Sedimentology* 3, 89–108.

Allen, J.R.L. (1985a) *Principles of Physical Sedimentation*. George Allen & Unwin: London.

Allen, P.A. (1985b) Hummocky cross stratification is not produced purely under progressive gravity waves. *Nature* 313, 562–564.

Allen, P.A. & Pound, C.J. (1985) Conference report: storm sedimentation. *J. Geol. Soc. Lond.* 142, 411–412.

Allen, P.A. & Underhill, J.R. (1989) Swaley cross-stratification produced by unidirectional flows, Bencliff Grit (Upper Jurassic), Dorset, UK. *J. Geol. Soc. Lond.* 146, 241–252.

Arnott, R.W. & Southard, J.B. (1990) Exploratory flow-duct experiments on combined flow bed configurations and some implications for interpreting storm-event stratification. *J. Sedim. Petrol.* 60, 211–219.

Blatt, H., Middleton, G.V. & Murray, R. (1980) *Origin of Sedimentary Rocks*. Prentice-Hall Inc: New Jersey.

Bottjer, D.J., Droser, M.L. & Jablonski, D. (1984) Bathymetric trends in the history of trace fossils. In: *New Concepts in the Use of Biogenic Sedimentary Structures for Paleoenvironmental Interpretation* (Ed. D.J. Bottjer). Soc. Econ. Paleont. Miner. Pac. Sect. Guidebook 52, 57–65.

Bourgeois, J. (1980) A transgressive shelf sequence exhibiting hummocky cross-stratification: the Cape Sebastian Sandstone (Upper Cretaceous), southwestern Oregon. *J. Sedim. Petrol.* 50, 681–702.

Bourgeois, J. (1983) Hummocks — do they grow? *Bull. Am. Ass. Petrol. Geol.* 67, 428.

Brenchley, P.J. (1985) Storm influenced sandstone beds. *Mod. Geol.* 9, 369–396.

Brenchley, P.J. & Newall, G. (1982) Storm influenced inner-shelf sand lobes in the Caradoc (Ordovician) of Shropshire, England. *J. Sedim. Petrol.* 52, 1257–1269.

Brenchley, P.J., Newall, G. & Stanistreet, I.G. (1979) A storm surge origin for sandstone beds in an epicontinental platform sequence, Ordovician, Norway. *Sedim. Geol.* 22, 185–217.

Campbell, C.V. (1966) Truncated wave-ripple laminae. *J. Sedim. Petrol.* **36**, 825–828.

Cheel, R.J. (1991) Grain fabric in hummocky cross-stratified storm beds: genetic implications. *J. Sedim. Petrol.* **61**, 102–110.

Cheel, R.J. & Leckie, D.A. (1991) Conglomeratic storm beds: their characteristics and paleoprocess interpretation. *Geol. Ass. Can. Ann. Meet. Progr. Abstr.* **16**, A22.

Colon, J.A. (1966) Some aspects of Hurricane Carla (1961): Hurricane Symposium. *Am. Ass. Oceanogr. Publ.* **1**, 1–33.

Craft, R.J. & Bridge, J.S. (1987) Shallow-marine sedimentary processes in the Late Devonian Catskill Sea, New York State. *Geol. Soc. Am. Bull.* **98**, 338–355.

Dott, R.H. & Bourgeois, J. (1982) Hummocky stratification: significance of its variable bedding sequences. *Geol. Soc. Am. Bull.* **93**, 663–680.

Dott, R.H. & Bourgeois, J. (1983) Hummocky stratification: significance of its variable bedding sequences: reply. *Geol. Soc. Am. Bull.* **94**, 1249–1251.

Duke, W.L. (1985) Hummocky cross-stratification, tropical hurricanes, and intense winter storms. *Sedimentology*, **32**, 167–194.

Duke, W.L. (1985) Hummocky cross-stratification, tropical hurricanes, and intense winter storms: reply. *Sedimentology*, **34**, 344–359.

Duke, W.L. (1990). Geostrophic circulation or shallow marine turbidity currents? The dilemma of paleoflow patterns in storm-influenced prograding shoreline systems. *J. Sedim. Petrol.* **60**, 870–883.

Duke, W.L., Arnott, R.W.C. & Cheel, R.J. (1991) Shelf sandstones and hummocky cross-stratification: new insights on a stormy debate. *Geology* **19**: 625–628.

Ekdale, A.A., Bromley, R.G. & Pemberton, S.G. (1984) *Ichnology: Trace Fossils in Sedimentology and Stratigraphy.* Soc. Econ. Paleont. Miner. Short Course 15.

Eyles, N. & Clark, B.M. (1986) Significance of hummocky and swaley cross-stratification in late Pleistocene lacustrine sediments of the Ontario basin, Canada. *Geology* **14**, 679–682.

Fielding, C.R. (1989) Hummocky cross-stratification from the Boxvale Sandstone Member in the northern Surat Basin, Queensland. *Aust. J. Earth Sci.* **36**, 469–471.

Flierl, G.R. & Robinson, A.R. (1972) Deadly surges in the Bay of Bengal: dynamics and storm-tide tables. *Nature (London)* **239**, 213–215.

Frey, R.W. (1990) Trace fossils and hummocky cross-stratification, Upper Cretaceous of Utah. *Palaios* **5**, 203–218.

Frey, R.W. & Pemberton, S.G. (1984) Trace fossil facies models. In: *Facies Models*, 2nd edn (Ed. R.G. Walker). Geosci. Can. Repr. Ser. 1, 189–207.

Gadow, S. & Reineck, H.-E. (1969) Ablandiger sandtransport bei sturmfluten. *Senckenberg. Mar.* **1**, 63–78.

Goldring, R. & Langenstrassen, F. (1979) Open shelf and nearshore clastic facies in the Devonian. *Spec. Pap. Palaeont.* **23**, 81–97.

Greenwood, B. & Sherman, D.J. (1986) Hummocky cross-stratification in the surf zone: flow parameters and bedding genesis. *Sedimentology* **33**, 33–45.

Hamblin, A.P. & Walker, R.G. (1979) Storm dominated shallow marine deposits: the Fernie-Kootenay (Jurassic) transition, southern Rocky Mountains. *Can. J. Earth Sci.* **16**, 1673–1690.

Handford, C.R. (1986) Facies and bedding sequences in shelf-storm-deposited carbonates—Fayetteville Shale and Pitkin Limestone (Mississippian), Arkansas. *J. Sedim. Petrol.* **56**, 123–137.

Harms, J.C., Southard, J.B., Spearing, D.R. & Walker, R.G. (1975) *Depositional Environments as Interpreted from Primary Sedimentary Structures and Stratification Sequences.* Soc. Econ. Paleont. Miner. Short Course 2.

Harms, J.C., Southard, J.B. & Walker, R.G. (1982) *Structures and Sequences in Clastic Rocks.* Soc. Econ. Paleont. Miner. Short Course 9.

Hayes, M.O. (1976) *Hurricanes as Geologic Agents: Case Studies of Hurricanes Carla, 1961, and Cindy, 1963.* Texas Bur. Econ. Geol. Rep. Invest. 61.

Howard, J.D. (1971) Comparison of the beach-to-offshore sequence in modern and ancient sediments. In: *Recent Advances in Paleoecology and Ichnology: Short Course Lecture Notes* (Eds J.D. Howard, J.W. Valantine & J.E. Warme) pp. 148–183. American Geological Institute: Washington, DC.

Howard, J.D. (1972) Trace fossils as criteria for recognizing shorelines in the stratigraphic record. In: *Recognition of Ancient Sedimentary Environments* (Eds J.K. Rigby & W.K. Hamblin). Spec. Publ. Soc. Econ. Paleont. Miner. Tulsa 16, 215–255.

Hunter, R.E. & Clifton, H.E. (1982) Cyclic deposits and hummocky cross-stratification of probably storm origin in Upper Cretaceous rocks of the Cape Sebastian area, southwestern Oregon. *J. Sedim. Petrol.* **52**, 127–143.

Kreisa, R.D. (1981) Storm-generated sedimentary structures in subtidal marine facies with examples from the Middle and Upper Ordovician of southwestern Virginia. *J. Sedim. Petrol.* **51**, 823–848.

Leckie, D.A. (1988) Wave formed, coarse-grained ripples and their relationship to hummocky cross-stratification. *J. Sedim. Petrol.* **58**, 607–622.

Leckie, D.A. & Krystinik, L.F. (1989). Is there evidence for geostrophic currents preserved in the sedimentary record of inner to middle-shelf deposits? *J. Sedim. Petrol.* **59**, 862–870.

Leckie, D.A. & Walker, R.G. (1982) Storm- and tide-dominated shorelines, in Cretaceous Moosebar–Lower Gates interval — outcrop equivalents of Deep Basin gas trap in Western Canada. *Am. Ass. Petrol. Geol. Bull.* **66**, 138–157.

McBride, E.F. & Yeakel, L.S. (1963) Relationship between parting lineation and rock fabric. *J. Sedim. Petrol.* **33**, 779–782.

McCrory, V.L.C. & Walker, R.G. (1986) A storm and tidally-influenced prograding shoreline — Upper Cretaceous Milk River Formation of southern Alberta, Canada. *Sedimentology* 33, 47–60.

Morton, R.A. (1981) Formation of storm deposits by wind-forced currents in the Gulf of Mexico and North Sea. In: *Holocene Marine Sedimentation in the North Sea Basin* (Eds S.D. Nio, R.T.E. Shuettenhelm & T.C.E. Weering). Int. Ass. Sedim. Spec. Publ. 5, 385–396.

Myrow, P.M. & Southard, J.B. (1991) Combined flow model for vertical stratification sequences in shallow marine storm-deposited beds. *J. Sedim. Petrol.* 61, 202–210.

Nelson, C.H. (1982) Modern shallow-water graded sand layers from storm surges, Bering Shelf: a mimic of Bouma sequences and turbidite systems. *J. Sedim. Petrol.* 52, 537–546.

Nøttvedt, A. & Kreisa, R.D. (1987) Model for the combined-flow origin of hummocky cross-stratification. *Geology* 15, 357–361.

Pantin, H.M. (1979) Interaction between velocity and effective density in turbidity flow: phase-plane analysis, with criterion for autosuspension. *Mar. Geol.* 31, 59–99.

Parker, G. (1982) Conditions for the ignition of catastrophically erosive turbidity currents. *Mar. Geol.* 46, 307–322.

Plint, A.G. & Walker, R.G. (1987) Cardium Formation 8. Facies and environments of the Cardium shoreline and coastal plain in the Kakwa field and adjacent areas, northwestern Alberta. *Bull. Can. Petrol. Geol.* 35, 48–64.

Prave, A.R. & Duke, W.L. (1990) Small-scale hummocky cross-stratification in turbidites: a form of antidune stratification? *Sedimentology* 37, 531–539.

Rosenthal, L.R.P. & Walker, R.G. (1987) Lateral and vertical facies sequences in the Upper Cretaceous Chungo Member, Wapiabi Formation, southern Alberta. *Can. J. Earth Sci.* 24, 771–783.

Rust, B.R. & Gibling, D.A. (1990) Three-dimensional antidunes as HCS mimics in a fluvial sandstone: the Pennsylvanian South Bar Formation near Sydney, Nova Scotia. *J. Sedim. Petrol.* 60, 540–548.

Seilacher, A. (1976) Bathymetry of trace fossils. *Mar. Geol.* 5, 413–428.

Sherman, D.J. & Greenwood, B. (1989) Hummocky cross-stratification and post-vortex ripples: length scales and hydraulic analysis. *Sedimentology* 36, 981–986.

Snedden, J.W., Nummedal, D. & Amos, A.F. (1988) Storm and fairweather combined flow on the central Texas continental shelf. *J. Sedim. Petrol.* 58, 580–595.

Swift, D.J.P. & Niedoroda, A.W. (1985) Fluid and sediment dynamics on continental shelves. In: *Shelf Sands and Sandstone Reservoirs* (Eds R.W. Tillman, D.J.P. Swift, & R.C. Walker) Soc. Econ. Paleont. Miner. Short Course Notes 13, 47–135.

Swift, D.J.P., Figueiredo, A.G., Freeland, G.L. & Oertel, G.F. (1983) Hummocky cross-stratification and megaripples: a geological double standard? *J. Sedim. Petrol.* 53, 1295–1317.

Swift, D.J.P., Han, H. & Vincent, C.E. (1986) Fluid process and sea floor response on a modern storm-dominated shelf: middle Atlantic shelf of North America: Part 1, the storm-current regime. In: *Shelf Sands and Sandstones* (Eds R.J. Knight & J.R. McLean). Can. Soc. Petrol. Geol. Mem. 11, 99–119.

Swift, D.J.P., Hudelson, P.M., Brenner, R.L. & Thompson, P. (1987) Shelf construction in a foreland basin: storm beds, shelf sandstones, and shelf-slope depositional sequences in the Upper Cretaceous Mesaverde Group, Book Cliffs, Utah. *Sedimentology* 34, 423–457.

Tillman, R.W. (1985) A spectrum of shelf sands and sandstones. In: *Shelf Sands and Sandstone Reservoirs* (Eds R.W. Tillman, D.J.P. Swift & R.G. Walker). Soc. Econ. Paleont. Miner. Short Course Notes 13, 1–46.

Vossler, S.M. & Pemberton, S.G. (1989) Ichnology and paleoecology of offshore siliciclastic deposits in the Cardium Formation (Turonian, Alberta, Canada). *Palaeogeogr. Palaeoclim. Palaeoecol.* 74, 217–239.

Walker, R.G. (1984) Shelf and shallow marine sands. In: *Facies Models*, 2nd edn (Ed. R.G. Walker). Geosci. Can. Repr. Ser. 1, 141–170.

Walker, R.G. (1985) Geological evidence for storm transportation and deposition on ancient shelves. In: *Shelf Sands and Sandstone Reservoirs* (Eds R.W. Tillman, D.J.P. Swift & R.G. Walker). Soc. Econ. Paleont. Miner. Short Course Notes 13, 243–302.

Walker, R.G., Duke, W.L. & Leckie, D.A. (1983) Hummocky stratification: significance of its variable bedding sequences: discussion. *Geol. Soc. Am. Bull.* 94, 1245–1249.

8   An introduction to estuarine lithosomes and their controls

JOHN R.L. ALLEN

Introduction

An estuary is transitional between a river and the shallow waters of a marine shelf or platform. It is a generally narrowing, elongated inlet reaching across a coastal plain or inward along a river valley as far as the upper limit of tidal rise, and in which tidal currents and interactions between river and salt-water have a dominating role. Estuaries are thus connected with fluvial and a range of marine environments. An estuarine lithosome (Krumbein & Sloss, 1963) is a volume of sediment of estuarine origin which interfingers with adjacent masses of a different character and origin, for example, fluvial, aeolian and shallow-marine. Dyer (1973), McLusky (1981), Pethick (1984), and Nichols & Biggs (1985) give introductions to the morphology, ecology and hydraulics of contemporary estuaries. The nature of estuarine deposits — the key to the successful interpretation of the rock record — is outlined by Clifton (1982) and by Nichols & Biggs (1985). Saltmarshes, a major component of estuaries, are reviewed by Frey & Basan (1978) and Long & Mason (1983).

Here I attempt to point to what we now know and understand about a variable, complex and much-studied sedimentary environment and its deposits. In a review as brief as this, I have thought it necessary to concentrate on the physical aspects of the estuarine environment, which define the framework within which operate geochemical and biological processes, deserving of separate extended treatment. The role of boundary conditions will be examined, followed by examples of estuaries of different classes and their deposits.

Boundary conditions

Geological context

The shape adopted by an estuary largely depends on its geological context. *Alluvial* or *coastal-plain estu-aries*, found today in areas of extensive Quaternary and especially post-glacial sedimentation, tend to a funnel or trumpet shape, typified by an exponential upstream decline in both flow width and depth (Wright *et al.*, 1973). This geometry is essentially a reflection of the interaction between the grades of sediment available and the hydraulic forces exerted by waves and particularly the tide. At the other extreme lie *rock-bound estuaries*, bordered by steep rock cliffs or bluffs of partly lithified sediment. Estuaries of this kind arise in regions of long-interrupted sedimentation, and their form depends on the sequence and structure of the confining resistant materials, rather than an estuarine process. These estuaries do not narrow regularly upstream, but are of uneven width and may locally change direction abruptly. Many estuaries, however, represent an intermediate state, the bedrock constraining their shape only to a limited degree.

The geological context of estuaries has implications for the sourcing of their sediments. Most suspended fine sediment reaching estuaries is introduced by rivers, but some can be derived from marginal cliffs or bluffs, from the adjoining coast outside the estuaries, or from the offshore sea-bed where exposed to wave or tidal scour, provided that suitable lithologies are present. Some, and in certain instances most, of the coarse sediment present in estuaries is introduced by rivers, with marginal cliffs or bluffs affording possible local sources, but in many cases is supplied largely from offshore. Sand in the Severn Estuary, for example, has a minera-logical signature (Barrie, 1980, 1981) suggesting supply from reworked glacial and fluvioglacial deposits on the sea-bed to the west.

Movement of sea-level

The rate and sense of movement of relative sea-level, by either providing or denying space for sediment accumulation, and by modifying the tidal and wave regimes, can profoundly influence the character and behaviour of estuaries and, consequently, the nature of the sedimentary sequences formed in them.

The movement of relative sea-level observed at a coastal site is the sum of many factors and there can be no support for a globally uniform behaviour. As regards south-west Britain in the post-glacial period (Heyworth & Kidson, 1982), an area where the Earth's crust appears to have bulged upward under the weight of nearby Pleistocene ice, the trend has been monotonically upward, at first rapidly but then more gradually (Fig. 8.1a). Calculations suggest that the tidal range in the Severn Estuary in this region may be gradually increasing as a result of the growing water depth (Tooley, 1985; see also Woodworth *et al.*, 1991). The behaviour of sea-level over broadly the same period in the Bay of Fundy is very different (Fig. 8.1b), the tidal range rapidly increasing from < 2 m during the period of fall to the order of 15 m today (Dalrymple *et al.*, 1990). In coastal Virginia, far to the south (Nichols, 1972), relative sea-level appears to have been moving steadily upward (Fig. 8.1c). A different pattern again is obtained from the Northern Territory of Australia (Woodroffe *et al.*, 1989) where, after a rapid ascent, relative sea-level stabilized about 6000 conventional radiocarbon years ago (Fig. 8.1d).

Tidal regime: water characteristics

The tidal regime, with its several higher-frequency periodicities (Pugh, 1987), is undoubtedly the single most important factor shaping estuaries and estuarine lithosomes. It controls the mixing between seawater and river water in an estuary, the power of the tidal streams, the transport of sediment within the system, and the character (especially the structural features) of the deposits accumulating there.

The tidal regime at most locations is *semidiurnal* (Fig. 8.2a), the height of the tide, and the velocity of the tidal currents, varying approximately twice-daily (period 12.43 h). Successive high waters and successive low waters normally show a zig-zag variation in height over time, called the *diurnal inequality*. *Spring tides*, when tidal height and range are at their greatest, occur fortnightly, roughly when the Moon is either new or full. These tides alternate with *neap tides*, when the height and range are least. The alternation of neap and spring tides defines the *spring–neap cycle*. Twice a year, at about the time of the equinoxes, the spring tides attain their greatest height and range. They are weakest, how-

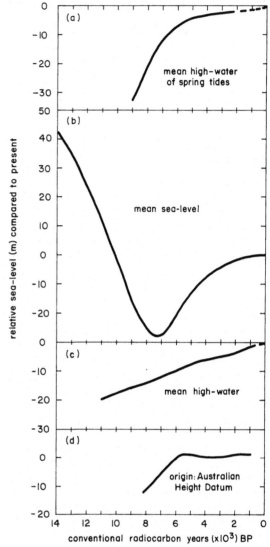

Fig. 8.1 Sea-level movements over the later Quaternary in representative estuaries (error bands not shown). (a) Bristol Channel and Severn Estuary, south-west Britain (from Heyworth & Kidson, 1982). (b) Chignecto Bay (Bay of Fundy), eastern Canada (from Dalrymple *et al.*, 1990). (c) Chesapeake Bay region, east coast of USA (from Nichols, 1972). (d) South Alligator Estuary, Northern Territory, Australia (from Woodroffe *et al.*, 1989). Each plot gives the indicative sea-level or origin stated (in the case of the South Alligator Estuary, the organic material dated comes from mangrove muds formed around high tide level).

ever, near the summer and winter solstices. Together these trends constitute an important *semiannual cycle*.

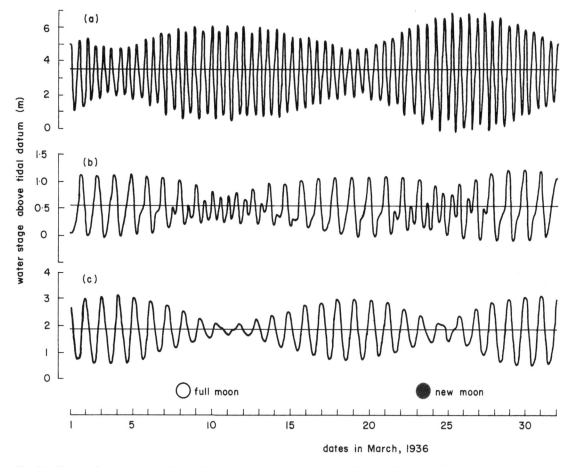

Fig. 8.2 Tidal regimes: (a) semi-diurnal (Immingham, eastern England); (b) mixed (Manila, Phillipines); and (c) diurnal (Do-Son, North Vietnam); illustrated by predicted tidal curves for March, 1936.

A *diurnal* tidal regime prevails at some coastal locations (Fig. 8.2c). Only one high water and one low water occur daily, and the fortnightly cycle, revealing an extreme variation in tidal height and range, varies with the phases of the moon in a different manner to the semidurnal regime (see Fig. 8.2a).

A *mixed* diurnal–semidurnal tidal regime is detectable at many coastal localities (Fig. 8.2b). Two more or less equal high waters and two more or less similar low waters are seen daily during parts of the fortnightly cycle, but at other times the high tides and low tides are either each very unequal or only one high water and one low water are evident in a day. The resulting complicated pattern of water movements depends on the relative strength of the semidiurnal and diurnal influences.

Another useful way of classifying tidal regimes is in terms of the average tidal range. Davies (1964) recognized *microtidal* (range < 2 m), *mesotidal* (range 2–4 m) and *macrotidal* (range > 4 m) regimes, and Kirby (1989) considered it useful to identify a *hypertidal* (range > 6 m) category. Microtidal regimes typify oceanic coasts, where little amplification of the small ocean tide can be expected from a narrow continental shelf. Mesotidal regimes imply a moderate to substantial amplification of the ocean tide and tend to occur where the shelf is broad. Macrotidal regimes typify broad and restricted shelf seas, such as the north-west European shelf with its irregular coastline and large and small islands.

Tidal range is a major influence on the degree of the mixing of river and saltwater in estuaries (Dyer,

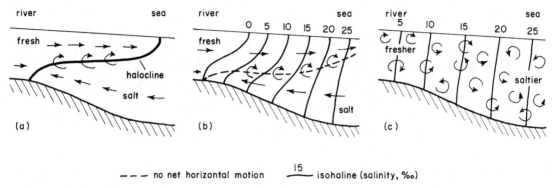

Fig. 8.3 Schematic salinity distributions in: (a) salt-wedge estuaries; (b) partially-mixed estuaries; and (c) well-mixed estuaries.

1973). In *salt-wedge estuaries* (Fig. 8.3a), normally microtidal, the tidal currents are comparatively weak, and the dense seawater underlies the less dense river water as a tapering wedge bounded above by a narrow mixing zone of sharp salinity change (halocline). The mixing induces a weak flow of saltwater into the estuary, in the opposite direction to the outward flow of freshwater. Suspended fine sediment tends to be concentrated around the null point between the opposing flows, creating a *turbidity maximum*. However, turbidity maxima may be present only during spring tides (e.g. Vale & Sundby, 1987). A *partially-mixed estuary* (Fig. 8.3b), most of which are either mesotidal or weakly macrotidal, occurs where the tidal streams are appreciable and create substantial shear and turbulence in the water body because of bed friction. In these, axial as well as vertical salinity gradients are apparent, and the halocline is thick and ill-defined. Strong saltwater inflows occur on account of the heightened mixing across the halocline. A turbidity maximum is normally present at the null point in the inner estuary, but another may be found in the middle or outer parts as a consequence of purely tidal effects. The position of the turbidity maximum follows an often lengthy excursion with the changing state of the tide. Macrotidal estuaries are typically of the *well-mixed* type (Fig. 8.3c). The intense shearing associated with the powerful tidal flows causes more or less complete mixing between river water and saltwater. There is a gradual axial salinity gradient but little or no vertical change in salt content; horizontally segregated zones of fresher and saltier water, in ebb- and flood-dominated channels, can be recognized across the width of such an estuary. Major turbidity max-

ima due to tidal effects may occur in the middle to inner parts of some well-mixed estuaries (e.g. Allen *et al.*, 1980) but in the outer parts of others (e.g. Kirby & Parker, 1983).

Hansen & Rattray (1966) quantified estuarine mixing with a simple plot (Fig. 8.4), in which ΔS is the difference in salinity between bed and surface at a cross-section of the estuary, S, the time-averaged salinity over that section, U, the time-averaged axial surface speed at the section, and U_r, the river discharge divided by the cross-sectional area. This versatile scheme can be used to emphasize that the degree of mixing varies with axial position and, choosing appropriate values of the time-dependent properties, with the state of the river and of the tide. The degree of the mixing normally improves as neap tides give place to spring tides (e.g. Allen *et al.*, 1980; Vale & Sundby, 1987), and as river floods subside (e.g. Hartwell, 1970; Nichols, 1977; Barua, 1990; Jay *et al.*, 1990).

Tidal regime: sediment movement

Because an estuary is closed at the head, the tidal wave within it is partially reflected, with the consequence that the tidal currents nominally are fastest at mid-tide and zero at each low water and high water (Fig. 8.5a,b). In practice, a number of effects (Dyer, 1973) combine to delay slack water a little, relative to low and high tides. This pattern of changing water stage and velocity predetermines the general vertical and horizontal distribution of sedimentary textures and structures in estuarine deposits (Fig. 8.5c,d) and, therefore, the general facies geometry of estuarine lithosomes. The estuarine biota is similarly zoned, in response to the

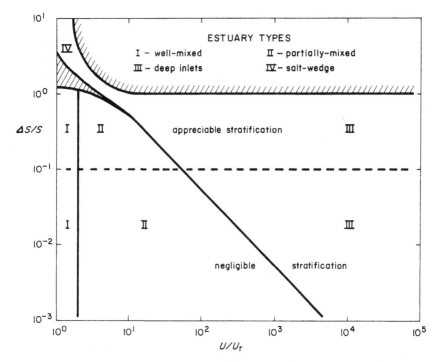

Fig. 8.4 Classification of estuaries by mixing characteristics on the basis of vertical salinity gradient and tidal relative to river flow (from Hansen & Rattray, 1966).

degree of emersion, sediment texture, and sediment mobility (McLusky, 1981; Long & Mason, 1983). Mud deposited from suspension at around the time of high-water slack necessarily accumulates high in the intertidal zone and has an excellent chance of preservation because, by the time the ebb stream is powerful enough to erode it, the tidal stage has dropped below the level of much of the deposit. Intermittent vigorous wave activity, however, may cause much resuspension throughout the high intertidal zone, and even lead to some net loss; heavy rainfall has a similar effect. The entrainment of sand calls for moderate currents, and the transport of this grade occurs only partly in suspension. Significant sand accumulations should therefore be restricted to the subtidal and lower intertidal zones. These sand bodies may be expected to display a range of surface and internal sedimentary structures (Southard & Boguchwal, 1990a,b), commensurate with the range of stream powers characteristic of sand-transporting tidal streams. Lag effects (Allen, 1982), however, will be severe; dunes and other large bedforms are commonly observed to survive the fall of the tide, when the tidal current declines

gradually toward zero as the water shallows. Gravel requires strong currents for its entrainment and transport, and is never suspended under tidal conditions. Where available, it should occur mainly subtidally, but could locally reach up to low in the intertidal zone, forming smooth sheets or dune fields. Theoretically, mud can also settle around the time of low-water slack, when a subtidal deposit may be expected (Fig. 8.5). Such accumulations appear to be infrequent, but may form by the repeated partial or complete decay, as during semi-diurnal and spring–neap cycles, of a turbidity maximum in an outer estuary where tidal streams begin to weaken (e.g. Kirby & Parker, 1983). Grossly, estuarine sequences should fine upward, but may locally display some upward coarsening in the basal part.

In detail, tidal sediments typically are marked by complex patterns of sedimentary textures and/or structures which record one, or a combination of, the several periodicities of the tidal regime, as expressed through the steeply non-linear increase in the power of tidal streams with increasing tidal range. These patterns involve such features as the

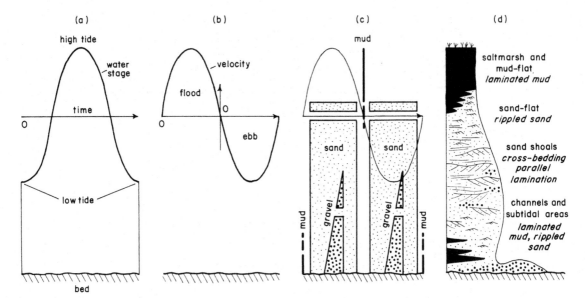

Fig. 8.5 Generalized estuarine facies model, developed from first principles. (a) Variation of water stage over a tidal cycle. (b) Variation of tidal velocity over a tidal cycle. (c) Time-height fields for the transport and deposition of mud, sand and gravel. (d) Vertical distribution of grain size and sedimentary structures. (It is assumed that the tidal flows at times attain upper-stage plane-bed conditions. In real estuaries, because of channel shifting, shoal deposits are likely in places to overlie erosional surfaces which may be channel-like.)

coupling of muddy with silty or sandy laminae, the partial to complete draping of internally cross-laminated ripple forms with mud, the coupling of groups of muddy laminae with groups of sandy ones, evidence of repeated current reversal, and the development within cross-bedding sets of intraset discontinuities (e.g. mud drapes, low-angle erosional discordances). Distinctive criteria for the recognition of ancient tidal sediments are thus created (Terwindt, 1981, 1988), and a basis is provided for the analysis of palaeotidal regimes (e.g. Klein, 1971; Clifton, 1983; Yang & Nio, 1985). Seasonal effects (Van den Berg, 1981), spring–neap cycles (Boersma & Terwindt, 1981; Van den Berg, 1982; De Mowbray & Visser, 1984), the diurnal inequality (Allen, 1985; De Boer et al., 1989), and individual semidiurnal tidal cycles are all potentially recognizable as lithological patterns in tidal deposits.

The tidal regime, interacting with estuary hypsometry, seems to determine whether an estuary either stores sediment made available from seaward, or whether it is either gradually emptying or serving to transmit fluvial coarse sediment from river to sea. Friedrichs et al. (1990) investigated this question for tide-dominated estuaries by means of a numerical model yielding a simple graphical criterion (Fig. 8.6). Here A is tidal amplitude, H, the mean channel depth, V_s, the volume of water stored intertidally at mean high water, and V_c, the volume of water stored in the tidal channels when the water level is at the height of mean sea-level. Comparatively deep estuaries with steep banks are likely to be ebb-dominated, that is, they are either emptying or acting as conduits, whereas shallow ones, with gentle, convex-up banks, are flood-dominated and serve as sediment attractors and stores. A sufficiently rapid rise in relative sea-level may convert a flood-dominated estuary into an ebb-dominated one. A sufficiently sharp fall, however, may change ebb dominance into flood dominance. Hence transgression need not favour the development and preservation of substantial estuarine lithosomes.

Climate

Climate determines many of the processes operating in the exposed intertidal zone of an estuary and, through its control of river hydrological regime, affects estuarine mixing (Fig. 8.4), causing

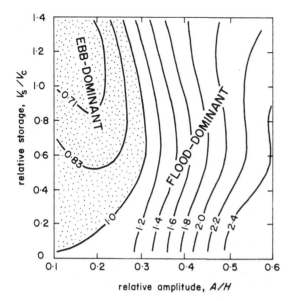

Fig. 8.6 Sediment transport conditions in estuaries, measured by the parameter ebbs/flood transport, as a function of tidal flow conditions and estuary shape (from Friedrichs *et al.*, 1990).

change on an annual to smaller scale. Estuaries in regions of strongly seasonal rainfall, for example, those of Sierra Leone (Tucker, 1973) and western India (Qasim & Sen Gupta, 1981), vary from well mixed or partially mixed during the generally long dry season to the salt-wedge type during the rains. A single, major flood event for a period converted the normally partially-mixed, microtidal Rappahannock Estuary, Virginia, into one of the salt-wedge type (Nichols, 1977). Tidal effects are also strongly suppressed during floods in the mesotidal Columbia Estuary, Oregon–Washington, when much suspended mud and fine sand flushes through the system (Jay *et al.*, 1990; Sherwood & Creager, 1990). A contrast is provided by the microtidal estuaries of Senegal–Gambia (Barusseau *et al.*, 1985), sited in a region of strongly seasonal rainfall and high temperatures. During the long dry season, these estuaries become hypersaline, expressed by the development of an 'inverse' salt wedge, as seawater (and with it sediment) is drawn in on the surface, and evaporated in the marginal uppermost intertidal and supratidal zones, and dense water accumulates at the bed. Similar evaporative losses may arise in the Australian Ord (Coleman & Wright, 1978) and Alligator (Woodroffe *et al.*, 1989) macrotidal estuaries, situated in a dry climate.

The intertidal zones of cold-climate estuaries are affected in various ways by the seasonal formation, movement and eventual melting of sea ice (Dionne, 1988). Coarse debris entrained by ice from beaches or cliffs can be drifted over saltmarshes and far out on to otherwise muddy intertidal flats, to the accompanying development of drag marks, miniature kettle holes, and a variety of other structures, all of which are potentially preservable. Estuaries freeze over, often for many months, where winters are severe. Intertidal sand shoals may then harden to a depth of some metres (e.g. Bartsch-Winkler & Ovenshine, 1984), a circumstance permitting blocks of ice-cemented sand to be incorporated after erosion and transport into the estuarine channel deposits on the remobilization of the shoals during the spring melt. Extensive deposits of fine sediment may form subtidally where there is a long-lived ice cover (D'Anglejan, 1980), because of the suppression of tide- and wave-related shear and turbulence, and such deposits may only be partly resuspended during the summer. There are indications that the intertidal sediments of subarctic estuaries contain very much less infauna (and are consequently less bioturbated) (D'Anglejan, 1980; Bartsch-Winkler & Ovenshine, 1984) than tidal mud flats in cool to temperate (e.g. Boyden & Little, 1973; Dalrymple *et al.*, 1990) and tropical (e.g. Tucker, 1973; Frey *et al.*, 1989) regions.

Sediments of microtidal estuaries

Little is known of the sediments of microtidal estuaries, although many microtidal bay and lagoonal deposits are reported.

Partly to largely rock-bound microtidal estuaries of small to moderate size are frequent along the Oregon–Washington coast, sited on an active continental margin (Kulm & Byrne, 1966, 1967; Peterson *et al.*, 1984). The fluvial supply of coarse sediment to these systems is important, but much sand enters from seaward, either at the estuary mouth, where tidal 'deltas' may occur, or indirectly by way of coastal aeolian dunes. The tidal currents are moderate to vigorous, and the estuarine sediments tend to fine inward along the axis of the system and vertically upward within the intertidal zone. Sands, coarsest in the outer estuary, dominate the mouth-bar deltas, channel floors and intertidal shoals, but give place to muddy sands and sandy muds on the high intertidal sand- and mud-flats. Because of the context, high intertidal flat deposits, and particularly saltmarsh sediments, are local in occurrence and volumetrically unimportant.

The Rappahannock Estuary, Virginia, one of the larger branches of Chesapeake Bay, is a rock-bound system bordered by bluffs of partly lithified mainly Tertiary and Quaternary sediments (Folger, 1972; Nichols, 1977). Peak tidal currents are moderate and decline from the mouth to the head of the estuary. The subtidal sediments are silts and, those of the intertidal flats, organic-rich muds, except where the bluffs provide a local source of sand. Organic-rich saltmarsh deposits occur locally as a fringe. The James River Estuary is also a comparatively deep, sheltered, partly rock-bound system toward the mouth of Chesapeake Bay (Nichols, 1972; Nichols *et al.*, 1991). The peak tidal currents tend to increase from the mouth to the head of the estuary, whereas the sediments fine from sands near the mouth to sandy to silty muds toward the head, where there is a turbidity maximum. The estuary is fringed by intertidal mud-flats and salt-marshes.

By contrast, the microtidal estuaries of the Senegal–Gambia coast have a coastal-plain setting in a hot, dry region of strongly seasonal rainfall (Barusseau *et al.*, 1985). Much coarse sediment is supplied from offshore by the moderate to vigorous tidal currents. The sands of the subtidal channels and intertidal sand shoals and sand-flats grade inward and upward into muddy sands and silty muds of high intertidal flats and mangrove swamps. As the dry season is 8–10 months long, the high intertidal and supratidal deposits may exhibit evaporitic tendencies.

Sediments of mesotidal estuaries

The large number of mesotidal estuaries which have been described, offer considerable variety, depending partly on context and climatic setting.

The Rupert Bay Estuary is a shallow, rapidly prograding system located in subarctic Canada within a Quaternary coastal plain where relative sea-level is falling by approximately 0.01 m annually (D'Anglejan, 1980). The tidal currents are vigorous, peaking at about $1.5 \, \mathrm{m \, s^{-1}}$, but the sediment chiefly reaching the estuary is a slightly sandy silt. This is finest grained on the subtidal floors of the shallow estuarine channels, becoming slightly coarser upward on the intertidal flats and salt-marshes fed and drained by a dendritic network of gullies. The intertidal deposits are rich in ice-

rafted coarse debris and in other indications of ice activity.

From the coastal plain on the eastern seaboard of the USA come sedimentological descriptions of the Saco, Maine (Farrell, 1970), the Essex (Boothroyd & Hubbard, 1974, 1975), the Merrimack (Hartwell, 1970) and the Parker (DaBoll, 1969; Boothroyd & Hubbard, 1974, 1975), Massachusetts, and the Assabaw (Greer, 1975), Georgia, estuaries. Vigorous tidal streams typify these systems; the currents in the Parker Estuary weaken toward the head. The main sand shoals occur around the mouths of the Saco and Merrimack estuaries, but in the Parker & Essex are distributed along with either ebb- or flood-dominated channels throughout most of the system. In all four, dunes occur subtidally and on the intertidal sand shoals. Upward in the intertidal zone, and toward the margins of the estuaries, rippled sand-flats occur which grade still higher into rippled and bioturbated flats formed of sandy to silty mud. These pass up into the well-laminated deposits of organic-rich salt-marshes, which occur as a fringe in the Saco Estuary but are very extensive in the Essex, Merrimack and Parker systems. The vertical sequence of facies is comparatively simple, with some variation from mouth to head (Fig. 8.7). The Assabaw lies on the open coast to the south-west and presents deposits similar to those in the Parker River Estuary and its companions. There are many parallels between these estuaries and a well-described Galician system (Vilas *et al.*, 1988). From Sapelo Island on the Georgia coast, Land & Hoyt (1966) described sandy tidal point bars such as may be expected in the upper reaches of small- to moderate-sized estuaries. Smith (1988) described some point bars from larger systems.

The Sixes River Estuary, a complex and variable rock-bound system on the Oregon coast (Boggs & Jones, 1976), experiences weak to moderate tidal currents. Dunes cover a tidal delta at the estuary mouth and a number of sand shoals in the lower estuary. A salt wedge occurs throughout the year but migrates seaward with increases in river flow. During summer, when there is little river discharge, fine sediment from seaward accumulates widely in the estuary, only to be flushed out in winter when the river floods.

The much larger Columbia River Estuary, on the Oregon–Washington border, is a mainly rock-bound system typified by very vigorous tidal currents and varying from the salt-wedge type to

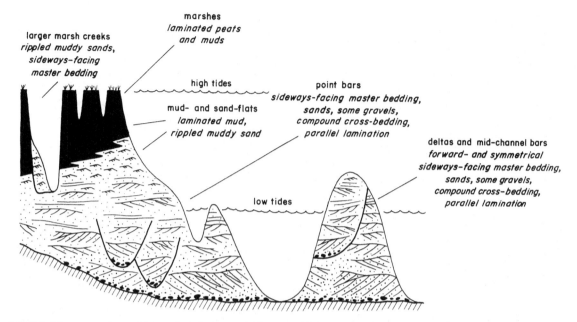

larger marsh creeks
rippled muddy sands,
sideways-facing
master bedding

marshes
laminated peats
and muds

high tides

mud- and sand-flats
laminated mud,
rippled muddy sand

point bars
sideways-facing master bedding,
sands, some gravels,
compound cross-bedding,
parallel lamination

deltas and mid-channel bars
forward- and symmetrical
sideways-facing master bedding,
sands, some gravels,
compound cross-bedding,
parallel lamination

low tides

Fig. 8.7 Diagrammatic, mainly vertical, facies model for mesotidal estuaries of the NE coast of the USA (from data in DaBoll, 1969; Farrell, 1970; Hartwell, 1970; Boothroyd & Hubbard, 1974, 1975).

well-mixed, depending on the state of the tide (mixed diurnal and semidiurnal regime) and the river inflow (Jay *et al.*, 1990; Sherwood & Creager, 1990; Sherwood *et al.*, 1990). Most of the sediment in the estuary has been supplied by the river, and there is a marked fining of the bed material from head to mouth. Because of the seasonal variation in wave activity and river discharge, there is a corresponding variation in the texture of the estuarine deposits. The exclusively subtidal floor of the outer estuary is marked by small reversing dunes. In the middle and upper reaches, there is a complex of shifting channels and elongated, intertidal sand shoals with large sand waves and dunes. Grain size fines upward on these shoals and into the bays on the margins of the estuary, where in some high intertidal flats and saltmarshes occur.

Southern England is an area where relative sea-level is moving upward at rates of up to a few millimetres annually. The Exe Estuary (Proctor, 1980; Thomas, 1980), deriving its sediment mainly from seaward, has a tidal delta just within the mouth, followed headward by dune-covered sand shoals and channels. Towards the margins, wide rippled sand-flats appear and there is fringe of prograding saltmarshes, representing what is left of very extensive marshes now largely reclaimed. By

contrast, the estuaries of the Crouch (Sheldon, 1968) and Medway (Kirby, 1990) are deep and mud-dominated and are experiencing rapid erosion of their intertidal flat and saltmarsh deposits.

The Scheldt Estuary in The Netherlands is well mixed and has vigorous tidal currents, peaking at about 2 m s^{-1}. The axial zone is a shifting complex of channels and subtidal to intertidal sand shoals covered with locally reversing dunes and sand waves (Boersma & Terwindt, 1981; Van den Berg, 1982; De Mowbray & Visser, 1984). These sands preserve an unusually clear and detailed structural record of the changing tidal regime. Seasonally banded sands and muds have accumulated subtidally in some channels (Van den Berg, 1981). Extensive minerogenic saltmarshes fed and drained by dendritic tidal creeks and gullies rim the estuary (Oenema & DeLaune, 1988). Van Straaten (1954, 1959) and Bouma (1963) described the possibly seasonal, upward-thinning, irregular laminations, much disturbed by roots, typical of such deposits (see also Frey & Basan, 1978).

The small estuaries of the Freetown coast, Sierra Leone, face the open Atlantic Ocean, experience a semidiurnal tidal regime, and contribute to a coastal plain typified by high temperatures and a strongly seasonal rainfall (Tucker, 1973). Mainly sand is driven into the estuaries from offshore during the dry season, whereas in the rains the rivers introduce mud and organic matter. Both the strength of the tidal currents and the grain size of the sediments tend to decline from the mouth to the head of each estuary. Dune-covered sand shoals and migrating channels with basal lag gravels mark the outer and axial zones of the estuaries. These give place toward the margins and the head to sandy or muddy intertidal flats with a variety of ripple and scour structures. Each estuary is fringed in the upper intertidal to low supratidal zones by mangrove swamps with silty to sandy muds. A dense and richly assorted infauna typifies the mangrove swamps and the sand and muddy intertidal flats. Because of rising relative sea-level, the channel and shoal sediments are prograding, at least partly erosively, over earlier intertidal flat and mangrove swamp deposits.

Sediments of macrotidal estuaries

Estuaries of this sort are reported from a variety of climatic and geological settings, and are generally either partially mixed or well mixed.

A subarctic climate, leading to a winter ice-cover lasting for several months, is the setting of the St Lawrence River Estuary, Canada (D'Anglejan, 1971, 1990; D'Anglejan & Smith, 1973; Dionne, 1988; Hamblin, 1989), and the rapidly prograding estuaries at the head of Cook Inlet, Alaska (Bartsch-Winkler & Ovenshine, 1984). Each system lies in a drowned, glaciated bedrock valley. The semidiurnal tidal regime of the St Lawrence Estuary creates tidal currents peaking at about 2.5 m s^{-1} and generates a substantial turbidity maximum in the middle estuary. The floor of the deep outer and middle estuary includes large, subtidal sand shoals covered in sand waves and dunes. A complex of braided intertidal sand shoals occurs in the upper estuary. Sandy to muddy intertidal flats, passing upward and laterally into saltmarshes, are present in the inner estuary and as a localized fringe, chiefly in bays lower down. Ice-transported coarse sediment and structures due to ice action typify the high intertidal flat and saltmarsh deposits. The tidal regime of the Alaskan estuaries is mixed diurnal–semidiurnal and exceptionally vigorous; the currents peak at about 4 m s^{-1}. Extensive braided sand-flats occur intertidally, revealing spreads of current ripples and, locally, dunes at low tide. Vertically within the flats, parallel lamination recording upper-stage plane-bed flows

gives place to parallel lamination interbedded with ripple cross-lamination and, finally, to ripple lamination. The most extensive saltmarshes lie at the heads of the estuaries.

Britain affords several examples of partly rock-bound, sand-dominated, macrotidal estuaries: in Wales, the Dovey (Haynes & Dobson, 1969), the Taf (Jago, 1980), and the Loughor (Elliott & Gardiner, 1981); and in Scotland, the Solway (Marshall, 1962; Bridges & Leeder, 1976; De Mowbray, 1983; Allen, 1989) and Tay (Buller, 1975; Buller & McManus, 1975; Buller et al., 1975; Green, 1975). Each obtains its coarse sediment chiefly from offshore, and each is characterized by vigorous tidal currents. Texturally, the sediments fine axially inward, outward and upward from the mouth and the channel deeps. Dune-covered channels and sand shoals in the outer and middle estuary give place headward and toward the margins to rippled and bioturbated sandy to muddy intertidal flats. The sediments are cross-bedded where dunes occur and mixed parallel-laminated and cross-laminated beneath the rippled flats. The muddier flats are underlain by interlaminated sands, silts and muds. Irregularly laminated saltmarsh deposits complete the intertidal sequence. The saltmarsh plants, by dampening the wave and tidal currents, are believed to trap and bind sediment; they commonly exhibit a strong zonation controlled by tidal immersion (Long & Mason, 1983). Supratidal aeolian dunes partly stabilized by vegetation overlook the entrances to the Dovey, Taf and Loughor estuaries.

By no means all British estuaries resemble these. Many macrotidal systems in south-west Britain are deep and cliff-bound, for example, the small estuary called Sandy Haven Pill (King, 1980). This sand-dominated estuary presents an axial zone of rippled sand-flats, combined with either mud-flats (outer estuary) or saltmarshes (inner estuary) which spread into gravel beaches below the cliffs. More muddy estuaries are well-represented by the Humber (Robinson, 1960; O'Connor, 1987; Armstrong, 1988; De Boer, 1988; Pethick, 1988), deriving most of its fine as well as coarse sediment from seaward.

The partly rock-bound and well-mixed Severn estuary in south-west Britain is a large retreating system noteworthy for its exceptional extreme tidal range (14.8 m, Avonmouth) and consequently very vigorous tidal flows (Stephens, 1986). Sediment supply and internal movement are complex (Collins, 1983, 1987). The sand-grade material, transported and recirculated within the system on mutually evasive paths (Harris & Collins, 1985), appears on mineralogical grounds (Barrie, 1980, 1981) to have been largely reworked from glacially introduced sources on the floor of the Celtic Sea and Bristol Channel to the west. Some fine sediment originates offshore (Murray & Hawkins, 1976; Murray, 1987), but most is introduced by the Severn and its tributaries (Allen, 1990, 1991a). A complex turbidity maximum, associated with an area of subtidal mud deposition, occurs at the entrance to the estuary (Kirby & Parker, 1983; Kirby, 1989). Sedimentologically, the estuary is divisible into three geographical parts (Hamilton, 1979; Allen, 1990). Because of the upward tendency (Fig. 8.1a) of relative sea-level (Heyworth & Kidson, 1982; Allen & Rae, 1988; Allen, 1991b), these sands are everywhere discordantly banked toward the margins of the estuary against the eroded edges of a thick sequence of estuarine silts and peats which record deposition high in the intertidal zone over the post-glacial period (Allen, 1990). The tidal currents are least vigorous in the outer system, where the axial zone is marked by a complex of channels and large intertidal sand shoals covered in reversing sand waves and dunes (Harris & Collins,

1985), presumably cross-bedded internally. The more vigorous tidal currents of the middle estuary shape smooth sand shoals typified internally by an alternation of parallel-lamination and ebb-dominated cross-lamination. Dunes appear locally only during the weaker tides. The inner estuary has a strongly meandering channel. Here internally cross-laminated and parallel-laminated sandy point bars overlying lag gravels grade up into coarsely laminated mud-flat and saltmarsh deposits. The saltmarshes of the estuary (Smith, 1979) are now largely reclaimed; at least in recent centuries, they have developed episodically in response apparently to a fluctuating wind-wave climate (Allen, 1987a; Allen & Rae, 1987). The saltmarsh and higher mud-flat deposits become extensively desiccated during the spring and summer (Allen, 1987b).

A tripartite zonation of estuarine environments comparable to that of the Severn Estuary is also recognized in the even more strongly macrotidal estuaries of the inner Bay of Fundy, Canada (Amos, 1978; Lambiase, 1980a,b; Yeo & Risk, 1981; Dalrymple *et al.*, 1990). Figure 8.8 gives a sedimentological model. Both the fine and the coarse sediments reaching these rapidly prograding, rock-

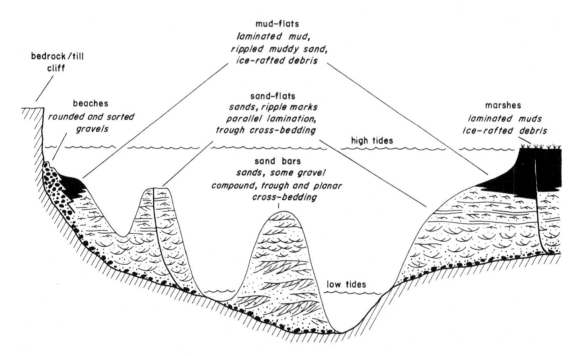

Fig. 8.8 Diagrammatic, mainly vertical, facies model for estuaries in the Bay of Fundy system, eastern Canada (from Lambiase, 1980a; Yeo & Risk, 1981; Dalrymple *et al.*, 1990).

bound, flood-dominated systems appear to be coming from offshore and from the margins of bedrock and till. The transport regimes are complex (Amos & Tee, 1989; Dalrymple et al., 1990). The sands in the outer zone of flow-parallel, linear bars with sand waves and dunes reveal several kinds of compound cross-bedding (Klein, 1970; Dalrymple, 1984; Dalrymple et al., 1990). The middle zone, of braided sand-flats, is typified by parallel-laminated and cross-laminated sands with cross-bedding where dunes are locally developed. On the margins, beaches of well-sorted gravel buttress the cliffs, and there are locally extensive sandy to muddy inter-tidal flats (variably bioturbated, sand–mud couplets) and saltmarshes (well-laminated sandy muds) (Yeo & Risk, 1981; Amos & Mosher, 1985; Gordon et al., 1985; Amos et al., 1988; Dalrymple et al., 1990).

The rock-bound Gironde Estuary (Allen et al., 1970, 1980; Gibbs et al., 1989) is long, deep and narrow, and a muddier macrotidal estuary than most of those described above. Tidal range and peak tidal currents increase toward the head; in the middle to inner estuary there is a turbidity maximum. The bed of the outer estuary is formed of mainly subtidal, flow-parallel sand shoals with some silt patches. These grade toward the margins into intertidal mud-flats and pockets of reedy saltmarsh. The marshes are more extensive along the margins of the middle estuary, the floor of which is formed almost wholly of mud. Dune-covered linear sand shoals, some grown upward to form narrow, stable islands, typify the inner estuary, but remain separated by muddy channels. Hence the overall textural sequence in the estuarine deposits is upward-fining only in the upper portion.

A good example of a tropical, mud-dominated estuary is provided by the South Alligator River in the coastal plain of the Australian Northern Territories (Woodroffe et al., 1989). Morphologically, this estuary consists of an outer funnel, followed headward by zones of sinuous and then cuspate meanders, and finally an irregular tidal channel. The channel sediments, typically interbedded sands and muds, are embedded in a thick sequence dominated by organic-rich muds formed below extensive mangrove forests which arose as sea-level stabilized about 6000 years ago (Fig. 8.1d). These overlie interbedded muds and sands of the estuarine funnel. The Ord River Estuary in Western Australia (Wright et al., 1973, 1975; Coleman & Wright, 1978) is similar morphologically to the South Alligator, but is much sandier, with many dune-covered, flow-parallel shoals. The climatic setting of each ensures that evaporite deposits overlie the mangrove-swamp muds, and that indicators of desiccation abound in the higher tidal sediments generally.

Conclusions

Just as estuaries range considerably in their geological and climatic settings, and in their tidal and river regimes, so estuarine lithosomes show no strong tendency to conform to one particular facies pattern.

The strongest tendencies, imposed by the daily or half-daily tide and by the presence of marginal constraints, are toward upward-fining and fining laterally toward the estuary margins. With marginal and upward-fining go changes in sedimentary structures indicative of weakening currents. There are none the less many exceptions to this tendency. In rock-bound systems, coarse-grained facies derived from marginal bluffs or cliffs can occur high in the tidal sequence. Muds are accumulating subtidally in the outer, deeper parts of a number of estuaries.

Most estuaries are sand-dominated, but a significant proportion are muddy. The coarse sediment in many cases originates to seaward, in which case the estuarine facies fine axially inward toward the head, as well as upward and marginally. River-dominated estuaries, and those subject to extreme flood events, however, receive most of their coarse sediment from the rivers, with the consequence that the facies fine from head to mouth.

The climatic setting, on the available evidence, manifests itself chiefly in the character of the uppermost intertidal sediments and in the density and diversity of the intertidal infauna. Seemingly, the infauna of cold-climate estuaries is sparse compared to that of temperate and tropical systems. Ice-rafted coarse debris and ice-related drag marks and scours are present among the high intertidal muds of cold-climate estuaries, whereas evaporite facies and profuse desiccation features mark this level in estuaries located in hot, dry regions.

Tidal range would appear to be expressed in estuarine lithosomes mainly in the stratigraphic scale of the subtidal–intertidal sand bodies, which lie parallel with flow, and in the vertical range within the intertidal muds in which evaporite and/or desiccation features occur. Both scales should increase with range. Because peak tidal currents grow with range, there is a tendency for sedimentary structures indicative of high stream powers to increase in prominence as the range increases from microtidal to macrotidal, under similar conditions of grain-size availability.

Estuarine lithosomes should be identifiable by the occurrence of features due to tidal action, the presence of flow-parallel sand bodies, and an overall geometry that is coast-normal and suggestive of inlet- or valley-filling. Their tendency toward upward-fining should distinguish them from deltaic deposits.

References

Allen, G.P., Castaing, P., Feral, A., Klingebiel, A. & Vigneaux, M. (1970) Contribution a l'étude des facies comblement et interpretation paleogeographique de l'évolution des milieux sédimentaires recents et actuels de l'éstuaire de la Gironde. *Bull. Inst. Géol. Bassin d'Aquitaine* **8**, 99–154.

Allen, G.P., Salomon, J.C., Bassoulet, P., Du Penhoat, Y. & de Grandpré, C. (1980) Effects of tides on mixing and suspended sediment transport in macrotidal estuaries. *Sediment. Geol.* **26**, 69–90.

Allen, J.R.L. (1982) *Sedimentary Structures*, Vol. 1. Elsevier: Amsterdam.

Allen, J.R.L. (1985) *Principles of Physical Sedimentology.* Allen & Unwin: London.

Allen, J.R.L. (1987a) Late Flandrian shoreline oscillations in the Severn Estuary: the Rumney Formation at its typesite (Cardiff area). *Phil. Trans. R. Soc.* **B315**, 157–184.

Allen, J.R.L. (1987b) Desiccation of mud in the temperate zone: studies from the Severn Estuary and eastern England. *Phil. Trans. R. Soc.* **B315**, 127–156.

Allen, J.R.L. (1989) Evolution of salt-marsh cliffs in muddy and sandy systems: a qualitative comparison of British west-coast estuaries. *Earth Surf. Processes Landforms* **14**, 85–92.

Allen, J.R.L. (1990) The Severn Estuary in southwest Britain: its retreat under marine transgression, and fine-sediment regime. *Sediment. Geol.* **66**, 13–28.

Allen, J.R.L. (1991a) Fine sediment and its sources, Severn Estuary and Bristol Channel, southwest Britain. *Sediment. Geol.* **75**, 57–65.

Allen, J.R.L. (1991b) Salt-marsh accretion and sea-level movement in the inner Severn Estuary, southwest Britain: the archaeological and historical contribution. *J. Geol. Soc. Lond.* **148**, 485–494.

Allen, J.R.L. & Rae, J.E. (1987) Late Flandrian shoreline oscillations in the Severn Estuary: a geomorphological and stratigraphical reconnaissance. *Phil. Trans. R. Soc.*, **B315**, 185–230.

Allen, J.R.L. & Rae, J.E. (1988) Vertical salt-marsh accretion since the Roman Period in the Severn Estuary, southwest Britain. *Mar. Geol.* **83**, 225–235.

Amos, C.L. (1978) The post glacial evolution of the Minas Basin, NS, a sedimentological interpretation. *J. Sedim. Petrol.* **48**, 965–982.

Amos, C.L. & Mosher, D.C. (1985) Erosion and deposition of fine-grained sediments of the Bay of Fundy. *Sedimentology* **32**, 815–832.

Amos, C.L. & Tee, K.T. (1989) Suspended sediment transport processes in Cumberland Basin, Bay of Fundy. *J. Geophys. Res.* **94**, 14407–14417.

Amos, C.L., Van Wagoner, N.A. & Daborn, G.R. (1988) The influence of subaerial exposure on the bulk properties of fine-grained intertidal sediment from Minas Basin, Bay of Fundy. *Estuar. Coast. Shelf Sci.* **27**, 1–13.

Armstrong, W. (1988) Life on the Humber. (A) Salt marshes. In: *A Dynamic Estuary: Man, Nature and the Humber* (Ed. N.V. Jones) pp. 46–57. Hull University Press: Hull.

Barrie, J.V. (1980) Heavy mineral distribution in bottom sediments of the Bristol Channel, UK. *Estuar. Coast. Mar. Sci.* **11**, 369–381.

Barrie, J.V. (1981) Hydrodynamic factors controlling the distribution of heavy minerals. *Estuar. Coast. Shelf Sci.* **12**, 609–619.

Bartsch-Winkler, S. & Ovenshine, A.T. (1984) Macrotidal subarctic environment of Turnagain and Knik Arms, upper Cook Inlet, Alaska: sedimentology of the intertidal zone. *J. Sedim. Petrol.* **54**, 1221–1238.

Barua, D.K. (1990) Suspended sediment movement in the estuary of the Ganges–Brahmaputra–Meghana river system. *Mar. Geol.* **91**, 243–253.

Barusseau, J.P., Diop, E.H.S. & Saos, J.L. (1985) Evidence of dynamic reversal in tropical estuaries, geomorphological and sedimentological consequences (Salem and Casamance Rivers, Senegal). *Sedimentology* **32**, 543–552.

Boersma, J.R. & Terwindt, J.H.J. (1981) Neap–spring tidal sequences of intertidal shoal deposits in a mesotidal estuary. *Sedimentology* **28**, 151–170.

Boggs, S. & Jones, C.A. (1976) Seasonal reversal of floodtide dominant sediment transport in a small Oregon estuary. *Bull. Geol. Soc. Am.* **87**, 419–426.

Boothroyd, J.C. & Hubbard, D.K. (1974) *Bed Form Development and Distribution Pattern, Parker and Essex Estuaries, Massachusetts.* US Army Corps of Engineers, Coastal Engineering Research Center, Miscellaneous Papers, 1–74.

Boothroyd, J.C. & Hubbard, D.K. (1975) Genesis of bedforms in mesotidal estuaries. In: *Estuarine Research*, Vol. 2 (Ed. L.E. Cronin) pp. 217–234. Academic Press: New York.

Bouma, A.J. (1963) A graphic presentation of the facies model of salt marsh deposits. *Sedimentology* **2**, 122–129.

Boyden, C.R. & Little, C. (1973) Faunal distributions in soft sediments of the Severn Estuary. *Estuar. Coast. Mar. Sci.* **1**, 203–223.

Bridges, P.H. & Leeder, M.R. (1976) Sedimentary model for intertidal mudflat channels, with examples from the Solway Firth, Scotland. *Sedimentology* 23, 533–552.

Buller, A.T. (1975) Sediments of the Tay Estuary. II. Formation of ephemeral zones of high suspended sediment concentrations. *Proc. R. Soc. Edin.* **B75**, 65–89.

Buller, A.T. & McManus, J. (1975) Sediments of the Tay Estuary. I. Bottom sediments of the upper and middle reaches. *Proc. R. Soc. Edin.* **B75**, 41–64.

Buller, A.T., Green, C.D. & McManus, J. (1975) Dynamics and sedimentation: the Tay in comparison with other estuaries. In: *Neashore Sediment Dynamics and Sedimentation* (Eds J. Hails & A. Carr) pp. 201–249. J. Wiley & Sons: Chichester.

Clifton, H.E. (1982) Estuarine deposits. In: *Sandstone Depositional Environments* (Eds P. Scholle & D. Spearing). Am. Ass. Petrol. Geol., Tulsa, pp. 179–189.

Clifton, H.E. (1983) Discrimination between subtidal and intertidal facies in Pleistocene deposits, Willapa Bay, Washington. *J. Sedim. Petrol.* 53, 353–369.

Coleman, J.M. & Wright, L.D. (1978) Sedimentation in an arid macrotidal alluvial river system: Ord River, Western Australia. *J. Geol.* 86, 621–642.

Collins, M.B. (1983) Supply, distribution, and transport of suspended sediment in a macrotidal environment: Bristol Channel, UK. *Can. J. Fisher. Aqua. Sci.* 40 (Suppl. No. 1), 44–59.

Collins, M.B. (1987) Sediment transport in the Bristol Channel. *Proc. Geol. Ass.* 98, 367–383.

DaBoll, J.M. (1969) *Holocene Sediments of the Parker River Estuary, Massachusetts.* Contrib. No. 3-CRG. University of Massachusetts: Massachusetts.

Dalrymple, R.W. (1984) Morphology and internal structure of sandwaves in the Bay of Fundy. *Sedimentology* 31, 365–382.

Dalrymple, R.W., Knight, R.J., Zaitlin, B.A. & Middleton, G.V. (1990) Dynamics and facies model of a macrotidal sand-bar complex, Cobequid Bay–Salmon River Estuary (Bay of Fundy). *Sedimentology* 37, 577–612.

D'Anglejan, B.F. (1971) Submarine sand waves in the St Lawrence Estuary. *Can. J. Earth Sci.* 8, 1480–1486.

D'Anglejan, B.F. (1980) Effects of seasonal changes on the sedimentary regime of a subarctic estuary, Rupert Bay (Canada). *Sedim. Geol.* 26, 51–68.

D'Anglejan, B.F. (1990) Recent sediments and sediment transport processes in the St Lawrence Estuary. In: *Oceanography of a Large-scale Estuarine System: the St Lawrence* (Eds M.I. El-Sabh & N. Silverberg) pp. 109–129. Springer-Verlag: Berlin.

D'Anglejan, B.R. & Smith, E.C. (1973) Distribution, transport and composition of suspended matter in the St Lawrence Estuary. *Can. J. Earth Sci.* 10, 1380–1396.

Davies, J.L. (1964) A morphogenetic approach to world shorelines. *Zeit. Geomorph.* Suppl. 8, 127–142.

De Boer, G. (1988) History of the Humber coastline. In: *A Dynamic Estuary: Man, Nature and the Humber* (Ed. N.V. Jones) pp. 16–30. Hull University Press: Hull.

De Boer, P.L., Oost, A.P. & Visser, M.J. (1989) The diurnal inequality of the tide as a parameter for recognition of tidal influences. *J. Sedim. Petrol.* 59, 912–921.

De Mowbray, T. (1983) The genesis of lateral accretion deposits in recent intertidal mudflat channels, Solway Firth, Scotland. *Sedimentology* 30, 425–435.

De Mowbray, T. & Visser, M.J. (1984) Reactivation surfaces in subtidal channel deposits, Oosterschelde, southwest Netherlands. *J. Sedim. Petrol.* 54, 811–824.

Dionne, J.C. (1988) Characteristic feature of modern tidal flats in cold regions. In: *Tide-Influenced Sedimentary Environments* (Eds P.L. De Boer, A. Van Gelder & S.D. Nio) pp. 301–332. Riedel: Dordrecht.

Dyer, K.R. (1973) *Estuaries: a Physical Introduction,* 140 pp. J. Wiley & Sons: London.

Elliott, T. & Gardiner, A.R. (1981) Ripple, megaripple and sandwave bedforms in the macrotidal Loughor Estuary, South Wales, UK. *Spec. Publ. Int. Ass. Sediment.* 5, 51–64.

Farrell, S.C. (1970) *Sediment Distribution and Hydrodynamics, Saco and Scarboro Estuaries, Maine.* Contrib. No. 6-CRG. University of Massachusetts: Massachusetts.

Folger, D.W. (1972) *Characteristics of Estuarine Sediments of the United States.* U.S. Geol. Surv. Prof. Paper 742.

Frey, R.W. & Basan, P.B. (1978) Coastal salt marshes. In: *Coastal Sedimentary Environments* (Ed. R.A. Davis) pp. 101–169. Springer-Verlag: New York.

Frey, R.W., Howard, J.D., Han, S.-J. & Park, B.-K. (1989) Sediments and sedimentary sequences on a modern macrotidal flat, Inchon, Korea. *J. Sedim. Petrol.* 59, 28–44.

Friedrichs, C.T., Aubrey, D.G. & Speer, P.E. (1990) Impacts of relative sea-level rise on evolution of shallow estuaries. In: *Residual Currents and Long-term Transport* (Ed. R.T. Cheng) pp. 104–122. Springer-Verlag: New York.

Gibbs, R.J., Tshudy, D.M., Konwar, L. & Martin, J.M. (1989) Coagulation and transport of sediments in the Gironde Estuary. *Sedimentology* 36, 987–999.

Gordon, D.C., Crawford, P.J. & Desplanque, C. (1985) Observations on the ecological significance of salt marshes in the Cumberland Basin, a macrotidal estuary in the Bay of Fundy. *Estuar. Coast. Shelf Sci.* 20, 205–227.

Green, C.D. (1975) Sediments of the Tay Estuary. III. Sedimentological and faunal relationships on the southern shore at the entrance to the Tay. *Proc. R. Soc. Edin.* **B75**, 91–112.

Greer, S.A. (1975) Estuaries of the Georgia coast, USA: sedimentology and biology. III. Sandbody geometry and sedimentary facies at the estuary–marine transition zone, Ossabaw Sound, Georgia: a stratigraphic model. *Senckenbergiana Maritima* 7, 105–135.

Hamblin, P.F. (1989) Observations and model of sediment transport near the turbidity maximum of the upper St Lawrence River. *J. Geophys. Res.* **94**, 14419–14428.

Hamilton, D. (1979) The high-energy, sand and mud regime of the Severn Estuary, SW Britain. In: *Tidal Power and Estuary Management* (Eds R.T. Severn, D.L. Dineley & L.E. Hawker) pp. 162–172. Scientechnica: Bristol.

Hansen, D.V. & Rattray, M. (1966) New dimensions on estuary classification. *Limnol. Oceanogr.* **11**, 319–326.

Harris, P.T. & Collins, M.B. (1985) Bedform distribution and sediment transport paths in the Bristol Channel and Severn Estuary, UK. *Mar. Geol.* **62**, 153–166.

Hartwell, A.D. (1970) *Hydrology and Holocene Sedimentation of the Merrimack River Estuary, Massachusetts.* Contrib. No. 5-CRG. University of Massachusetts: Massachusetts.

Haynes, J. & Dobson, M. (1969) Physiography, foraminifera and sedimentation in the Dovey Estuary (Wales). *Geol. J.* **6**, 217–256.

Heyworth, A. & Kidson, C. (1982) Sea-level changes in southwest England and Wales. *Proc. Geol. Ass.* **93**, 91–111.

Jago, C.F. (1980) Contemporary accumulation of marine sand in a macrotidal estuary, southwest Wales. *Sediment. Geol.* **26**, 21–49.

Jay, D.A., Giese, B.S. & Sherwood, C.R. (1990) Energetics and sedimentary processes in the Columbia River estuary. *Progr. Oceanogr.* **25**, 157–174.

King, C.J.H. (1980) A small cliff-bound estuarine environment: Sandyhaven Pill in South Wales. *Sedimentology* **27**, 93–105.

Kirby, R. (1989) Sediment problems arising from barrage construction in high energy regions: an example of the Severn Estuary. In: *Third Conference on Tidal Power (Institution of Civil Engineers)* pp. 189–200. Thomas Telford: London.

Kirby, R. (1990) The sediment budget of the erosional zone of the Medway Estuary, Kent. *Proc. Geol. Ass.* **101**, 63–77.

Kirby, R. & Parker, W.R. (1983) Distribution and behaviour of fine sediment in the Severn Estuary and inner Bristol Channel, UK. *Can. J. Fisher. Aqua. Sci.* **40**, Suppl. 1, 83–95.

Klein, G. deV. (1970) Depositional and dispersal dynamics of intertidal sand bodies. *J. Sedim. Petrol.* **40**, 1095–1127.

Klein, G. deV. (1971) A sedimentary model for determining palaeotidal range. *Bull. Geol. Soc. Am.* **82**, 2585–2592.

Krumbein, W.C. & Sloss, L.L. (1963) *Stratigraphy and Sedimentation.* Freeman & Co.: San Francisco.

Kulm, L.D. & Byrne, J.V. (1966) Sedimentary response to hydrography in an Oregon estuary. *Mar. Geol.* **4**, 85–118.

Kulm, J.M. & Byrne, J.V. (1967) Sediments of Yaquina Bay, Oregon. In: *Estuaries* (Ed. G. Lauff). Am. Ass. Adv. Sci. Washington, DC, 226–238.

Lambiase, J.J. (1980a) Sediment dynamics in the macro-

tidal Avon River Estuary, Bay of Fundy, Nova Scotia. *Can. J. Earth Sci.* **17**, 1628–1641.

Lambiase, J.J. (1980b) Hydraulic control of grain-size distributions in a macrotidal estuary. *Sedimentology* **27**, 433–446.

Land, L.S. & Hoyt, J.E. (1966) Sedimentation in a meandering estuary. *Sedimentology* **6**, 191–207.

Long, S.P. & Mason, C.F. (1983) *Saltmarsh Ecology.* Blackie: Glasgow.

McLusky, D.S. (1981) *The Estuarine Ecosystem.* Blackie: Glasgow.

Marshall, D.R. (1962) The morphology of the upper Solway salt marshes. *Scott. Geogr. Mag.* **78**, 81–99.

Murray, J.W. (1987) Biogenic indications of suspended sediment transport in marginal marine environments: quantitative examples from southwest Britain. *J. Geol. Soc. Lond.* **144**, 127–133.

Murray, J.W. & Hawkins, A.B. (1976) Sediment transport in the Severn Estuary during the past 8000 to 9000 years. *J. Geol. Soc. Lond.* **132**, 385–398.

Nichols, M.M. (1972) Sediments of the James River, Virginia. *Mem. Geol. Soc. Am.* **133**, 169–219.

Nichols, M.M. (1977) Response and recovery of an estuary following a river flood. *J. Sedim. Petrol.* **47**, 1171–1186.

Nichols, M.M. & Biggs, R.B. (1985) Estuaries. In: *Coastal Sedimentary Environments* (Ed. R.A. Davis) pp. 77–186. Springer-Verlag: New York.

Nichols, M.M., Johnson, G.H. & Peebles, P.C. (1991) Modern sediments and facies model for a microtidal coastal plain estuary, the James Estuary, Virginia. *J. Sedim. Petrol.* **61**, 883–899.

O'Connor, B.A. (1987) Short and long term changes in estuary capacity. *J. Geol. Soc. Lond.* **144**, 187–195.

Oenema, O. & DeLaune, R.D. (1988) Acretion rates in salt marshes in the eastern Scheldt. *Estuar. Coast. Shelf Sci.* **26**, 379–394.

Peterson, C., Scheidegger, K., Komar, P. & Neim, W. (1984) Sediment composition and hydrography in six high-grade estuaries of the northwestern United States. *J. Sedim. Petrol.* **54**, 86–97.

Pethick, J.S. (1984) *An Introduction to Coastal Geomorphology.* Edward Arnold: London.

Pethick, J.S. (1988) Physical characteristics of the Humber. In: *A Dynamic Estuary: Man, Nature and the Humber* (Ed. V.N. Jones) pp. 31–45. Hull University Press: Hull.

Proctor, M.C.F. (1980) Vegetation and environment in the Exe Estuary. In: *Essays on the Exe Estuary* (Ed. G.T. Boalch) pp. 117–134. The Devonshire Association: Exeter.

Pugh, D.T. (1987) *Tides, Surges and Mean Sea-level.* John Wiley & Sons: Chichester.

Qasim, S.Z. & Sen Gupta, R. (1981) Environmental characteristics of the Mandovi-Zuari Estuary system of Goa. *Estuar. Coast. Shelf Sci.* **13**, 557–578.

Robinson, A.H.W. (1960) Ebb-flood channel systems in

sandy bays and estuaries. *Geography,* **45,** 183–199.

Sheldon, R.W. (1968) Sedimentation in the estuary of the River Crouch, Essex, England. *Limnol. Oceanogr.* **13,** 72–83.

Sherwood, C.R. & Creager, J.S. (1990) Sedimentary geology of the Columbia River Estuary. *Progr. Oceanogr.* **25,** 15–79.

Sherwood, C.R., Jay, D.A., Harvey, R.B., Hamilton, P. & Simenstad, C.A. (1990) Historical changes in the Colombia River Estuary. *Progr. Oceanogr.* **25,** 299–352.

Smith, D.G. (1988) Modern point bar deposits analogous to the Athabasca oil sands, Canada. In: *Tide-influenced Sedimentary Environments and Facies* (Eds P.L. De Boer, A. Van Gelder & S.D. Nio) pp. 417–432. Riedel: Dordrecht.

Smith, L. (1979) *A Survey of Salt Marshes in the Severn Estuary.* Nature Conservancy Council: London.

Southard, J.B. & Boguchwal, L.A. (1990a) Bed configurations in steady unidirectional flows. Part 2. Synthesis of flume data. *J. Sedim. Petrol.* **60,** 658–679.

Southard, J.B. & Boguchwal, L.A. (1990b) Bed configurations in steady flows. Part 3. Effects of temperature and gravity. *J. Sedim. Petrol.* **60,** 680–686.

Stephens, C.V. (1986) A three-dimensional model for tides and salinity in the Bristol Channel. *Cont. Shelf Res.* **6,** 531–560.

Terwindt, J.H.J. (1981) Origin and sequences of sedimentary structures in inshore mesotidal deposits of the North Sea. *Spec. Publ. Int. Ass. Sediment.* **5,** 4–26.

Terwindt, J.H.J. (1988) Palaeo-tidal reconstructions of inshore tidal depositional environments. In: *Tide-influenced Sedimentary Environments* (Eds P.L. De Boer, A. Van Gelder & S.D. Nio) pp. 233–263. Riedel: Dordrecht.

Thomas, J.M. (1980) Sediments and sediment transport in the Exe Estuary. In: *Essays on the Exe Estuary* (Ed. G.T. Boalch) pp. 73–87. The Devonshire Association: Exeter.

Tooley, M.J. (1985) Sea levels. *Progr. Phys. Geogr.* **9,** 113–120.

Tucker, M.E. (1973) The sedimentary environments of tropical African estuaries: Freetown Peninsula, Sierra Leone. *Geol. Mijn.* **52,** 203–215.

Vale, C. & Sundby, B. (1987) Suspended sediment fluctuations in the Tagus Estuary on semi-diurnal and fort-nightly time scales. *Estuar. Coast. Shelf Sci.* **25,** 495–508.

Van den Berg, J.H. (1981) Rhythmic seasonal layering in a mesotidal channel fill sequence, Oosterschelde, The Netherlands. *Spec. Publ. Int. Ass. Sediment.* **5,** 147–159.

Van den Berg, J.H. (1982) Migration of large-scale bedforms and preservation of cross-bedded sets in highly accretionary parts of tidal channels in the Ooster-schelde, SW Netherlands. *Geol. Mijn.* **61,** 253–263.

Van Straaten, L.M.J.U. (1954) Composition and structure of recent marine sediments in the Netherlands. *Leidse Geol. Mededelinger* **19,** 1–110.

Van Straaten, L.M.J.U. (1959) Minor structures of some recent littoral and neritic sediments. *Geol. Mijn.* **21,** 197–216.

Vilas, F., Sopena, A., Rey, L., Ramos, A., Nombela, M.A. & Arche, A. (1988) The Corrubedo tidal flat, Galicia, NW Spain: sedimentary processes and facies. In: *Tide-influenced Sedimentary Environments* (Eds P.L. De Boer, A. Van Gelder & S.D. Nio) pp. 183–200. Riedel: Dordrecht.

Woodroffe, C.D., Chappell, J., Thom, B.G. & Wallensky, E. (1989) Depositional model of a macrotidal estuary and floodplain, South Alligator River, Northern Australia. *Sedimentology* **36,** 737–756.

Woodworth, P.L., Shaw, S.M. & Blackman, D.L. (1991) Secular trends in mean tidal range around the British Isles and along the adjacent European coastline. *Geophys. J. Int.* **104,** 593–609.

Wright, L.D., Coleman, J.M. & Thom, B.G. (1973) Processes of channel development: Cambridge Gulf–Ord River delta, Western Australia. *J. Geol.* **81,** 15–41.

Wright, L.D., Coleman, J.M. & Thom, B.G. (1975) Sediment transport and deposition in a macrotidal river channel: Ord River, Western Australia. In: *Estuarine Research,* Vol. 2 (Ed. L.E. Cronin) pp. 309–321. Academic Press, New York.

Yang, C.-S. & Nio, S.-D. (1985) The estimation of palaeo-ohydrodynamic processes from subtidal deposits using time series analysis methods. *Sedimentology* **32,** 41–57.

Yeo, R.K. & Risk, M.J. (1981) The sedimentology, stratigraphy, and preservation of intertidal deposits in the Minas Basin system, Bay of Fundy. *J. Sedim. Petrol.* **51,** 245–260.

Index

Page numbers in *italics* refer to figures and tables.